Sustainable Material, Design, and Process

This text emphasizes the importance of sustainable material, design, and manufacturing processes, and how the needs are changing day by day. It comprehensively covers important topics including material recycling, optimal utilization of resources, green materials, biocomposites, clean and green synthesis, stable material properties, utilization of renewable energy sources, ergonomic design, and sustainable design. The text examines the design process, manufacturing, and upscaling of next-generation materials and their application in diverse industries. The text is primarily written for graduate students and academic researchers in the fields of manufacturing engineering, materials science, mechanical engineering, and environmental engineering.

- Presents an in-depth understanding of the progress of the need for new innovative and next-generation materials.
- Discusses biocomposites and green materials for eco-friendly products in a comprehensive manner.
- Explores recycling techniques of materials for sustainable manufacturing.
- Presents conceptual framework of sustainable product development.
- Covers important topics such as process optimization, renewable energy, and 3D printing in detail.

The text discusses the designing process of these new materials, manufacturing, and upscaling of these materials along with their selection for industrial applications. It further focuses on improving surface homogeneity in nanoparticle scattering during dip coating for stable and efficient wettability during oil/water separation. It will serve as an ideal reference text for graduate students and academic researchers in the fields of manufacturing engineering, materials science, mechanical engineering, and environmental engineering.

Sustainable Material, Design, and Process

Edited by
Ravi Kant
Hema Gurung
Shashikant Yadav

CRC Press
Taylor & Francis Group
Boca Raton London New York

CRC Press is an imprint of the
Taylor & Francis Group, an **informa** business

First edition published 2024
by CRC Press
6000 Broken Sound Parkway NW, Suite 300, Boca Raton, FL 33487-2742

and by CRC Press
4 Park Square, Milton Park, Abingdon, Oxon, OX14 4RN

CRC Press is an imprint of Taylor & Francis Group, LLC

© 2024 selection and editorial matter Ravi Kant, Hema Gurung and Shashikant Yadav; individual chapters, the contributors

British Library Cataloguing-in-Publication Data
A catalogue record for this book is available from the British Library

Library of Congress Cataloging-in-Publication Data
Names: Kant, Ravi (Professor of Chemistry), editor. |
Gurung, Hema, editor. | Yadav, Shashikant, editor.
Title: Sustainable material, design and process / edited by Ravi Kant,
Hema Gurung, and Shashikant Yadav.
Description: First edition. | Boca Raton : Taylor and Francis, 2024. |
Includes bibliographical references and index.
Identifiers: LCCN 2023009642 (print) | LCCN 2023009643 (ebook) |
ISBN 9781032150505 (hbk) | ISBN 9781032150529 (pbk) | ISBN 9781003242291 (ebk)
Subjects: LCSH: Manufacturing processes. | Green products--Materials. | Product design.
Classification: LCC TS183 .S875 2024 (print) | LCC TS183 (ebook) |
DDC 670.286--dc23/eng/20230420
LC record available at https://lccn.loc.gov/2023009642
LC ebook record available at https://lccn.loc.gov/2023009643

ISBN: 978-1-032-15050-5 (hbk)
ISBN: 978-1-032-15052-9 (pbk)
ISBN: 978-1-003-24229-1 (ebk)

DOI: 10.1201/9781003242291

Typeset in Times
by MPS Limited, Dehradun

Contents

Preface

Manufacturing is one of the most important sectors for the overall development of a nation and society. It helps in bringing independence, security, and power due to independent self-reliant foreign policy, and a stable social and political system by enhancing comfortability, prosperity and security among the people. However, the manufacturing sector is facing various issues including high setup and production costs, the need for skilled manpower, harmful emissions, demand for customized production, high energy requirement, low efficiency, disposal of waste, etc. These issues are not only because of manufacturing processes but also due to design and materials. These issues are general obstacles to the growth of the manufacturing sector in any country especially for the developing countries. Sustainability in the manufacturing sector is important for low-cost eco-friendly production for the overall development of society and the country.

Sustainable materials, design and processes are needed to reduce environmental impact and foster a sense of connection between people and nature. Therefore, the requirements for sustainable materials, designs and processes are essential to meet current environmental, economic and cultural challenges. This book bridges the knowledge gap by providing a way to develop the professional skills needed to produce sustainable materials, designs and processes. It encourages designers to regain control over development and manufacturing processes, reiterating the importance of sustainable materials, designs and processes.

Sustainable technologies are those that use less energy, deplete natural resources and do not pollute the environment directly or indirectly, and recycle or reuse the product once its purpose is served. It also emphasizes applying professional expertise to directly or indirectly impact the environment. This book covers a wide range of ideas ranging from innovative, socially acceptable avenues for environmental protection to environmentally friendly technological advancements.

Chapter 1 presents an investigation of coating angles, coating pressures and coating distances during jet spray coating on the membrane for oil-water separation applications. These parameters were varied until optimal levels were reached to establish the membrane with fewer clusters on the surface. Chapter 2 focuses on the synthesis of biodegradable packaging material using cellulose extracted from agricultural waste rice straw, in combination with polyethylene glycol (PEG) as plasticizer along with tartaric acid (TA) and citric acid (CA) serving as both cross-linkers and antioxidant agents. The novel PEG-reinforced cellulose films with strong interaction between them could enrich the fabrication of high-performance biodegradable packaging films.

Chapter 3 presents the limitations of the conventional tricycle rickshaw design and reveals that traditional rickshaws use age-old technology and poor mechanical design. It presents a low-cost improved design for manually driven tricycles which can be used in local transportation for personal, recreational, and small-scale commercial uses. Chapter 4 includes a review of biodegradable materials and their applications to enhance the utilisation of these sustainable bio-based materials.

Chapter 5 discusses that the artificially roughened solar air heater absorbers possess a high thermal efficiency. The studies on fluid flow and heat transmission are carried out in a channel of discretized broken V-type baffle and continuous V-type baffle. The use of the periodic V-shaped baffles caused turbulence with a considerable heat transfer rate & pressure fall. These discrete V- type baffles gave the highest thermo- hydraulic performance.

Chapter 6 presents the design and manufacturing of tools, jigs and fixtures for using a milling machine as a Friction Stir Welding (FSW) setup. The developed setup is used to join aluminium and copper alloy plates. The investigation delivered good weld joints for similar and dissimilar materials with good weld quality and excellent weld strength, and frictionless surface finishes with no gap created between the welded plates. Chapter 7 discusses the numerical analysis of alternative designs and configurations of radiator fins in order to have a comparative data analysis on the basis of efficiency and effectiveness. The fin designs are studied using COMSOL 5.5 for geometric configurations of four different modifications with rectangular fins, additional squared fins, circular fins and diametric change of radiator tubes. Chapter 8 presents the usage of Minimum Quantity Lubrication (MQL) during the mechanical machining of metallic materials for saving cost and protecting the environment. Further, the advantages of using vegetable oils instead of mineral oils are explored in terms of ecological and economic benefits.

Chapter 9 presents the use of Green Cutting Fluid (GCF) by mixing nontoxic emulsifiers and coconut oil for cutting EN31 steel, titanium alloy Ti6Al4V, and aluminium alloy AA5052. The concentration of GCF is optimized considering the surface topography and roughness. It is concluded from the research that the GCF is a practical, long-term replacement for mineral oil-based cutting fluids due to superior wetting qualities and advantages for the environment. Chapter 10 aims to identify ergonomic issues associated with the use of analogue micrometers and its association with an error that is made during measurements. A quantitative as well as qualitative approach is employed. Study findings reveal that physical and visual ergonomic usability problems occur in micrometers due to design aspects, nature of work, environment and behavioral reasons.

The chapters highlight the limitations, propose solutions and report relevant results for improvement in various designs, manufacturing processes and materials. They also highlight the research gaps and directions for future work. The chapters are a good combination of review articles and original research works. It will help readers to gain broader and deep knowledge in the field of sustainability in manufacturing. The content of this book will be useful for academicians, researchers and practicing engineers who are looking for sustainability in materials, design and manufacturing processes. We welcome the feedback from the readers about this book.

Editors
Ravi Kant, Hema Gurung, and Shashikant Yadav

Editors

Dr. Ravi Kant is an Assistant Professor in the Department of Mechanical Engineering at Indian Institute of Technology Ropar, India. He earned his Bachelor degree in Mechanical Engineering from Maharshi Dayanand University, Rohtak (Haryana, India). He completed his M. Tech. from Department of Mechanical Engineering at Indian Institute of Technology (IIT) Guwahati, India with specialization in "Computer Assisted Manufacturing". He worked on the Investigation on Formability of Adhesively Bonded Sheets during his M. Tech. project. He also earned his doctorate from IIT Guwahati in the field of laser forming process. His research interests include Laser transmission welding; Hybrid machining; Laser forming; Cold spray coatings; Additive manufacturing, Hybrid joining and Sustainable materials. He has completed many research projects and consultancy works in these research areas. He has contributed around 80 research articles in peer-reviewed journals, conferences, and edited books. He has edited one book titled "Simulations for design and manufacturing" published by Springer Singapore. He has also guest-edited five special issues in reputed journals. He has developed and taught advanced courses like modern manufacturing processes; sustainability science and technology; analysis of casting, forming and joining processes; advanced welding technology; micromanufacturing; manufacturing, etc. He has also conducted various international conferences, workshops, symposiums, colloquiums, and faculty development programs in the field of advanced manufacturing technology.

Dr. Hema Gurung is an independent researcher in the field of Robotics. She earned her bachelor's degree from Kalyani University (West Bengal, India) in Electronics and Instrumentation Engineering. She completed her M. Tech. in Mechatronics from IIEST (Indian Institute of Engineering and Science Technology, Shibpur, India) and earned her Ph.D. from the Department of Mechanical Engineering at Indian Institute of Technology Guwahati, India in the field of Robotics. She had worked at Hanbat National University, South Korea and Thapar University, India. Her research work includes experimental analysis, numerical simulation, optimization, control systems, state estimator, robotics, smart sensors and actuators. She has published a good number of research articles in international journals, conferences and edited books. Currently, she is working at the application of smart sensors and actuators in Industry 4.0.

Dr. Shashikant Yadav is an Assistant Professor in the Department of Chemical Engineering at Dr B R Ambedkar National Institute of Technology Jalandhar, India. He earned his Bachelor's degree in Chemical Engineering from Deenbandhu Chhotu Ram University of Science and Technology, Murthal Sonipat (Haryana, India). He completed his M. Tech. from the Department of Chemical Engineering at Indian

Institute of Technology (IIT) Roorkee, India with specialization in "Computer Aided Process Plant Design". He completed his Ph.D. from Department of Chemical Engineering at Indian Institute of Technology (IIT) Bombay, India on the topic "Carbon sequestration by carbon dioxide in mineral slurries and industrial waste". His research interests include CO_2 Sequestration, Multiphase Reactions, Mineral Carbonation, Waste Utilization and Management, Analytical, Numerical and Soft–Computing Modeling and Optimization of Reaction Processes, Social and Economic Engineering, Edible films and coatings.

Contributors

Alugongo A. A.
Vaal University of Technology
Vanderbijlpark 1900
Private Bag X021
South Africa

Arjun S
Sai Vidya Institute of Technology
Bangalore, India

Badiuddin Aliya F.
Indian Institute of Science
Bengaluru, India

Baruah Mrinmoy G.
Tezpur University
Assam, India

Beniwal Preeti
Panjab University
Chandigarh, India

Chatha Sukhpal Singh
Yadavindra College of Engineering
Punjabi University Campus
Talwandi Sabo, Punjab, India

Das Harjyoti
Tezpur University
Assam, India

Dheeraj KG
Sai Vidya Institute of Technology
Bangalore, India

Dutta Partha Pratim
Tezpur University
Assam, India

Dutta Polash P.
Tezpur University
Assam, India

Edachery Vimal
Indian Institute of Science
Bengaluru, India

Gokul Sarvepally S
Sai Vidya Institute of Technology
Bangalore, India

Kailas Satish V.
Indian Institute of Science
Bengaluru, India

Maome T.G.
Vaal University of Technology
Vanderbijlpark 1900
Private Bag X021
South Africa

Mishra Anjan Kumar
Parala Maharaja Engineering College
Odisha, India

Padia Neel P
Indian Institute of Information
    Technology Design and
    Manufacturing
Madhya Pradesh, India

Pal Amrit
GZS Campus College of Engineering
    and Technology
MRSPTU
Punjab, India

Palai Siddhant
Parala Maharaja Engineering College
Odisha, India

Pandit Sangeeta
Indian Institute of Information
    Technology Design and
    Manufacturing
Madhya Pradesh, India

Pattnayak Sritam
Parala Maharaja Engineering College
Odisha, India

Pawar Amol S
Sai Vidya Institute of Technology
Bangalore, India

Rao Boddepalli Durga
Parala Maharaja Engineering College
Odisha, India

Rishikesh M R
Sai Vidya Institute of Technology
Bangalore, India

Sahoo Rashmi Rekha
Parala Maharaja Engineering College
Odisha, India

Sharma Malvika
Panjab University
Chandigarh, India

Sharma Shivashree
Dibrugarh University
Assam, India

Sidhu Hazoor Singh
Yadavindra College of Engineering
Punjabi University Campus
Talwandi Sabo, Punjab, India

Sindhu Ravi
Indian Institute of Science
Bengaluru, India

Sob P. B.
Mount Vernon Nazarene University
Ohio, USA

Subramanya Raghavendra
Sai Vidya Institute of Technology
Bangalore, India

Suvin P. S.
National Institute of Technology
  Karnataka
Suratkal, India

Tengen T. B.
Vaal University of Technology
Vanderbijlpark 1900
Private Bag X021
South Africa

Tomy Abel
Indian Institute of Science
Bengaluru, India

Toor Amrit Pal
Panjab University
Chandigarh, India

Virdi Roshan Lal
Punjabi University
Punjab, India

Yadav Saroj
Dibrugarh University
Assam, India

1 Modelling the Ceramic Membrane Surface Properties to Minimize Clusters during Nanoparticle Coating by Jet-Spray for Efficient Wettability during Oil-Water Separation

T.G. Maome

Department of Industrial Engineering, Operations
Management and Mechanical Engineering, Faculty of
Engineering and Technology, Vaal University of Technology,
Vanderbijlpark 1900, Private Bag X021, South Africa

P. B. Sob

Mount Vernon Nazarene University Ohio-USA

A. A. Alugongo and T. B. Tengen

Department of Industrial Engineering, Operations
Management and Mechanical Engineering, Faculty of
Engineering and Technology, Vaal University of Technology,
Vanderbijlpark 1900, Private Bag X021, South Africa

CONTENTS

DOI: 10.1201/9781003242291-1

1

1.1   INTRODUCTION

For years now, ceramic materials have been used as the main membrane materials, coated with nanoparticles to give the desired wettability properties for oil-water separation [1–4]. There are other materials used in membrane wettability design for oil-water separation. Such materials are glass, polymers, clay, textiles and sediments [5–7]. These materials have different wetting properties when coated with nanoparticles, which affect their wettability [6]. Ceramic membrane has been the best material used in membrane technology [7–9]. The main advantage of using ceramic materials is due to their enhanced thermal stability, chemical stability and mechanical properties when compared to other materials used in membrane technology [1–3,10–13]. The main problem faced in the manufacturing of hydrophilic/oleophobic or hydrophobic/oleophilic nanostructured ceramic membranes is due to coating techniques that are employed during coating [3,14–16]. These techniques are suspension plasma spray coating, spin coating, dip coating, electrochemical deposition and jet-spray coating [9,15,16]. These techniques give different effects during coating [13,15,16]. Suspension plasma spraying is a type of coating method which is widely used for coating the membrane surface with nanoparticles. It has the capability of coating complex shapes [17–19]. The process is based on supplying the feedstock powder material into a very high-temperature plasma jet where it is heated and accelerated with high-velocity flow. However, this method is very expensive to run due to that it requires only argon to be used as plasma gas, hence it was not recommended for use in this research. The spin coating technique is a well-known technique in many industrial sectors for coating flat surfaces [20]. In this process, a coating solution is dropped on a vacuum pump, normally on a fixed rotational speed between 400 to 1200 rpm for a certain spinning time (normally from 30 s to 10 min). However, this technique is only limited to coat flat substrates and not complex substrates [20], hence it was not recommended for use in this research. The dip coating method is a preferred method used for complex geometrical membranes [20]. Such membranes have closely spaced struts which can prompt capillary forces [20]. However, this results in channel/cell clogging [20], hence it was not recommended for use in this research. The electrochemical deposition has emerged as a competitive technique to coat large-area surfaces during nanoparticle coating [21]. The process is based on immersing electrodes (as membranes) in an aqueous solution [21]. The electrodes are then energized with a direct current (DC) to allow for electrolysis [21]. The deposition time may differ, depending on the required surface properties [21]. The coated membrane is then immersed in alcohol (mostly ethanol) then dried out to achieve the required nanostructured membrane [21]. However, this results in clusters developed on the membrane. To combat these clusters, depositioning time and voltage must be increased until the required surface properties are achieved [21]. This technique is energy-intensive, hence it was not recommended for use in this research. The jet-spray coating technique is a type of coating used to fabricate appropriate nanostructured membranes [22,23]. In ceramics, it is used in fabricating microfiltration (MF), ultrafiltration (UF) and nanofiltration (NF) membranes [2,22]. The jet-spray coating has been the best strategy amongst other strategies due to that it offers the following: better scattering of nanoparticles on the membrane surface, optimal surface coverage during coating, high deposition speed of nanoparticles and it is flexible to cover any shape of the substrate [2,16–18,22], hence it was selected for use in this research. There are two main methods of producing nanostructured membranes using the jet-spray gun. It is either by high-pressure (HP) or low-pressure (LP) coating [17,18]. Several studies have been reported on cluster formation after the coating process during the manufacturing of nanostructured ceramic membranes [17–19,24,25]. These physical conditions that caused cluster formation are the aggregation of nanoparticles and different scattering effects of nanoparticles [18,19]. Clusters have a major effect on membrane performance [17–19,24]. They appear differently on ceramic membranes during HP and LP coating, which gives different effects on membrane wettability [17,18]. These clusters result in an unstable and uncontrollable separation efficiency of the membrane due to the different scattering effects of nanoparticles [26–31]. Different clusters are revealed from different coating rounds [17,19,24] therefore, the method of nanoparticle coating is vital in minimizing membrane surface clusters [17–19,24].

These methods of coating revealed different surface clusters which are discussed in the research. An approach has been proposed to minimize these clusters due to different physical conditions that affect the flow of nanoparticles on the ceramic surface during the coating process. It has been reported that more clusters are normally observed on LP coating than on HP coating when using jet-spray coating, hence the study focused more on HP coating. To minimize these clusters, it is important to model the parameters which will result in fewer and small size clusters on the membrane surface during nanoparticle coating. These parameters are coating force, coating pressure, coating angle and coating distance.

To achieve optimal membrane wettability, the membrane surface should have exceptional surface properties where the mixture of oil and water will be separated to a point where both clean oil and clean water are collected separately [8,9,32]. Wettability of the coated membrane is characterized by measuring contact angles on the membrane surface [33,34].

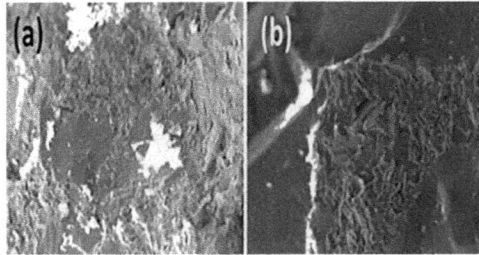

FIGURE 1.1   Shows SEM images of LP jet spray-coated membrane 1 (a) and HP jet spray-coated membrane 1 (b) [8,9,32].

It has been reported that HP and LP coating give different clustering on the membrane surface when using jet-spray coating [17]. With different coating rounds, clusters appear differently on the membrane. This is due to a different scattering effect of nanoparticles on the membrane surface during coating rounds [17]. Sob et al (2019) reported more and large clusters observed on LP jet-spray coating as compared to HP jet-spray coating as depicted in Figure 1.1(a) & (b).

In the current study, a theoretical model has been developed to minimize the formation of clusters during HP and LP coating on rough and smooth ceramic membrane surfaces. To design the ceramic membrane with fewer clusters developed on the surface, the following parameters were considered during HP and LP coating, coating force, coating pressure, coating distance and coating angle. For a rough membrane, the coating pressure was gradually increased to optimum levels during the coating rounds. This gives better viscous flow and scattering effect of nanoparticles during HP jet-coating. The number of HP coating rounds gave optimal membrane surface smoothness with fewer clusters developed on the membrane. For a smooth membrane, the coating pressure was maintained at optimum levels, but in a steady state during the coating rounds. This is to maintain better viscous flow and scattering effect of nanoparticles during HP jet-coating.

1.2   METHODOLOGY

Figure 1.2 demonstrates the process flow of ceramic membrane surface modification. It indicates some of the internal factors that affect material wettability during oil-water separation. It also indicates some of the external factors which will have an effect on material coating processes and their wettability during nanoparticle coating. In designing the ceramic nanostructured membrane, the following parameters were modelled for efficient wettability during oil-water separation: coating force, coating pressure, coating distance and coating angle. The model identified was tested using the concept of the stochastic approach. The model was tested for surface energy-driven separability. To achieve the research goal, the focus was on the characterization of the cluster sizes on the wettable nanostructured membrane. The approaches followed a sequence of coating nanoparticles on wettable membranes using a jet-spray gun and then tested the

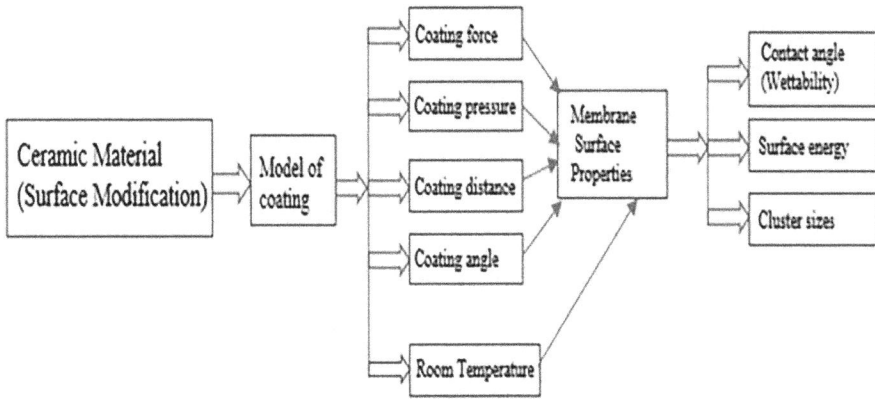

FIGURE 1.2 Modelling of surface energy-driven separability jet-spray gun during coating.

model for surface energy-driven separability. Therefore, the major experimental parameter variables were the nanoparticle coating, coating force, coating pressure, coating distance, coating angle and surface energy.

1.2.1 RESEARCH APPROACH

This project was approached in two stages. Stage one of this project focused on a theoretical framework based on ceramic membrane material modification of surface properties by nanoparticle coating and stage two was experimentation and validation through surface energy-driven separability. Our theoretically developed models of surface wettability (e.g., coating force, coating angle and coating distance) guided the manufacturing process and nanoparticle coating strategy of the wettable membrane surfaces. The ceramic membrane system was fabricated by using hydrophobic-oleophilic nano4stone nanoparticles containing *Fluorine S5* as the control element. The coating technique that was used in this project is the jet-spray technique. This technique has a high deposition rate and is flexible to reach any shape of the substrate [8,9]. It has unique properties such as thickness uniformity and offers high scattering effects across the surface [8,9].

Characterization was done using scanning electron microscopy (SEM), transmission electron microscopy (TEM), energy dispersion spectroscopy (EDS) and ImageJ particle analyser. This work focused on the effects of coating different rounds of LP and HP to identify the optimum coating pressure which will give a better scattering effect of nanoparticles thus reducing clusters on the wettable membrane. Since the morphology of membranes, nanoparticle sizes and scattering of nanoparticles on the membrane are naturally random, the concept of a stochastic approach was used. The stochastic approach promotes the analysis of spontaneous phenomena as reported by Sob *et al* (2020). The improved relationship between ceramic membrane surface properties was established. Such properties are surface smoothness, contact angle, surface tension and surface energy.

1.2.2 METHOD

1.2.2.1 Theoretical Modelling and Simulation Of Surface Energy-driven Separability

Figure 1.3(a) and Figure 1.4(a) show a jet-spray gun during high-pressure and low-pressure coating, used to produce nanostructured membranes for oil-water separation.

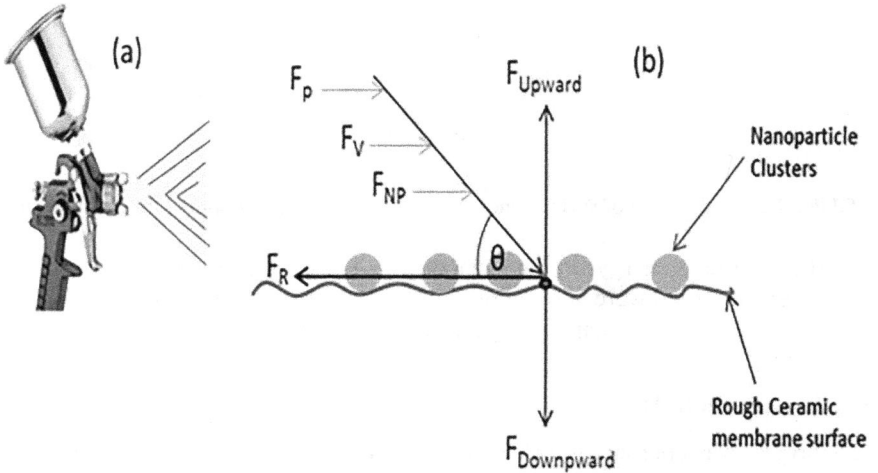

FIGURE 1.3 (a) Jet-spray gun during coating and (b) movement of nanoparticles on a rough membrane surface during coating.

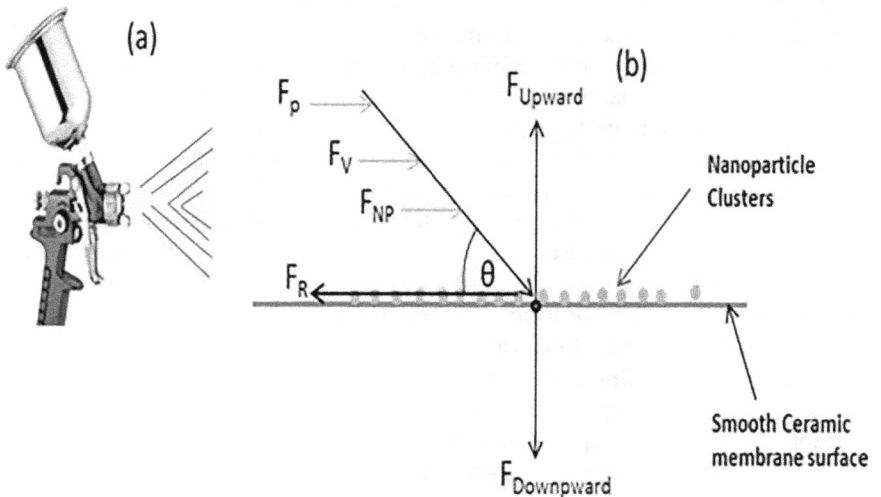

FIGURE 1.4 (a) Jet-spray gun during coating and (b) movement of nanoparticles on a smooth membrane surface during coating.

Impurities (which may be oil or water) must flow through the produced membrane surface. For the water-oil mixture to be effectively separated from the coated membrane surface, the effect of nanoparticles on the coated ceramic solid membrane surface must be considered. Nanoparticles coated on the ceramic membrane surface have different surface properties which affects the surface energy of the membrane. The roughness or smoothness of the ceramic membrane surface played a significant role in how water and oil flew on the membrane surface. Other forces such as the force of nanoparticles (F_{nano}) to lower the surface energy on the membrane surface, the force of viscosity ($F_{viscosity}$), the force on nanoparticles due to applied pressure from the spray gun ($F_{nano/pressure}$), the force on the solid wall due to nanoparticles (F_{down}) and the reaction force on the wall due nanoparticles (F_{upward}) are shown in Figure 1.3(b) and Figure 1.4(b).

Figure 1.3(a) shows a jet-spray gun during coating and Figure 1.3(b) depicts a rough ceramic membrane surface during coating rounds and the mobility of nanoparticles on the membrane surface which must be modelled and studied for the external and internal parameters that affect surface homogeneity. From Figure 1.3(b) the different forces that impacted surface homogeneity are greatly dependent on the pressure of the jet-spray gun, the distance of the jet-spray gun from the membrane during coating and the frictional resistance force (F_R) that impacted the force of nanoparticle (F_{nano}), the force of viscosity ($F_{viscosity}$), the force on nanoparticle to due applied pressure ($F_{nano/pressure}$), the force on a solid wall and nanoparticle ($F_{downward}$) and the force on wall and nanoparticle (F_{upward}). It was observed during the coating of the rough ceramic membrane surface that the frictional force must be high for the nanoparticles to move through the rough membrane surface during the coating process. Figure 1.4 revealed membrane coating for a smooth ceramic surface, however, the frictional forces were low when compared to a rough membrane surface during coating rounds in order to minimize the formation of clusters as shown in Figure 1.4(b)

Figure 1.4(a) shows a jet-spray coating on a smooth ceramic membrane surface and 4(b) depicts the mobility of nanoparticles in a smooth ceramic surface under the influence of low resistance force when compared with a rough membrane surface. Therefore, membrane frictional resistance forces are vital forces when minimizing the formation of clusters in ceramics membrane surface. The coating pressure and frictional resistance force must be maintained at optimal levels, but in a steady state to maintain proper movement and a better scattering effect of particles on the membrane surface. Before modelling the frictional resistance force and scattering effect that impact membrane surface clusters in a ceramic membrane, it is important to look at the total forces acting on a rough membrane and smooth membrane during ceramic membrane coating and is given as:

$$F_{Total} = F_{NPcos\ \theta} + F_{(Viscosity)} + F_{(Spray\ gun\ pressure)} + F_{(Downwards)}$$
$$- F_{(Frictional\ resistance)} \tag{1.1}$$

Equation (1.1) gives a total force that acts on a membrane surface during the coating process. The formation of membrane clusters is due to membrane surface roughness.

(a)

(b)

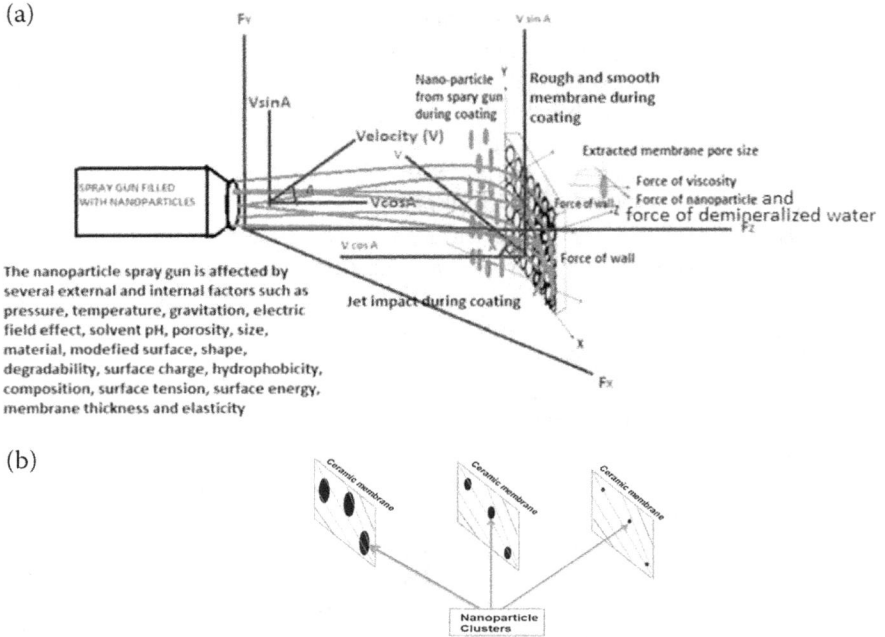

FIGURE 1.5 (a) Schematic of jet impact propulsion forces during membrane coating by jet-spray gun [8,32]. (b) Ceramic membrane surfaces with different cluster sizes.

This surface roughness is influenced by the frictional resistance force during coating. A rough membrane surface requires higher frictional force and a smooth membrane surface requires low frictional force for membrane clusters to be minimized. Therefore, there is a need to investigate the optimal frictional force in a rough and smooth membrane surface to minimize the surface clusters during coating. Figure 1.5 revealed the forces that are acting on a coated membrane surface during the coating process when using the jet-spray gun.

From Figure 1.5, the jet-spray propulsion during nanoparticle coating depends on the jet diameter, coating pressure or force, fluid viscosity, velocity, distance of the jet spray gun from the coating membrane and the coating angles of the ceramic membrane used in the coating. These parameters affect membrane surface clusters and must be modelled to minimize membrane clusters during coating. Since the motion of nanoparticles during coating is parabolic as shown in Figure 1.5(a), the jet-spray gun makes an angle θ with the horizontal and vertical component velocity at point C is given as $V \cos \theta \cos \theta$ and $V \sin \sin \theta$ as shown in Figure 1.6(a). If another point at D there are coordinates (x, y) before the nanoparticles strike the ceramic membrane and the particles travel at a given time (t) to hit the ceramic membrane surface as shown in Figure 1.5(a). The x and y components in terms of velocity, the angle of projection and time of projection are given as $x = V \cos \theta \cos \theta$ (t) and $y = V \sin \theta \sin \theta (t) - \frac{1}{2}gt^2$. From the x component of the velocity, t can be computed as $t = \frac{x}{V \cos \theta \cos \theta}$. Substituting t in y yielded:

FIGURE 1.6 (a) Crushed ceramic membrane (b) Ceramic membrane grains submerged in pre-clean detergent (c) Ceramic membrane grains dried up after cleaning (d) Coating the ceramic membrane using jet spray (e) Drying up ceramic membrane samples after coating (f) Parceled ceramic membrane samples for microscopic analysis (g) Ceramic membrane control sample (h) Nano4 Stone control sample.

$$y = x\frac{\sin\ \theta\ \sin\ \theta}{\cos\ \theta\ \cos\ \theta} - \frac{gx^2}{2V^2\cos\ \theta\ \cos\ \theta} \qquad (1.2)$$

From Equation (1.2) the impact of membrane clusters during jet-spray coating can be analysed based on the coating angles of the jet-spray gun θ, the speed of nanoparticles from the jet-spray gun V and the distance the jet-spray gun is kept away from the ceramic membrane during coating. The coating process revealed the flow of nanoparticles that can be related to flow in the open surface since the coating process was done in a controlled laboratory environment. The velocity from Figure 1.5(a) can be derived based on the physical reality during the coating process. The nanoparticle discharge by the jet spray as shown in Figure 1.5(a) would have covered an external distance S_3 while spending external time, t_3 as the nanoparticles flew from the jet-spray gun to the ceramic membrane being coated. To estimate these external characteristics, flow characteristics were measured from the jet-spray gun to the coated membrane as shown in Figure 1.5(a).

Since the nanoparticles hit the membrane surface after fleeing the S_3 at a time t_3, the impact of the jet-spray gun is felt on the membrane surface. This impact can be related to the mass flow rate on the membrane surface discharge by the jet-spray gun given as $M_3 = \rho A_3 V_3$ where ρ the density of nanoparticle, A_3 is the nanoparticle coating cross-sectional area of the surface and V_3 velocity of nanoparticles during coating. The average time t_3 spent by nanoparticles to get to the membrane surface was measured using a stopwatch. It was obtained by calculating the time it took the nanoparticle to flow from the jet-spray gun and when the nanoparticle struck the ceramic membrane during coating. This was done after repeated trials and by taking the average time t_3, and assuming that the average time flow at a constant speed V_3. The average distance travelled by the nanoparticle during jet-spray coating can be received from the speed formula $S_3 = V_3 t_3$. Therefore, the distance covered by nanoparticles during jet-spray coating is given as:

$$\Delta S = S_3 - S_1 = \frac{M_3 t_3}{\rho A_3} - S_1 \qquad (1.3)$$

The change of distance covered by nanoparticles during coating can be computed by looking at the specific capacity from the jet-spray gun given by assuming a limit of a function q which depends on a single point (x,y) at the injection point of the jet-spray gun given as the specific capacity q which is given as $q = \frac{\Delta Q}{\Delta S}$ where ΔQ define the nanoparticle flowing through the surface of the jet-spray gun during coating and Δs is the faction of the coated surface. The function of specific capacity during jet-spray coating depends on the angular distribution of the jet-spray gun during coating given as:

$$q_\theta = \frac{\partial Q}{\partial \theta} \qquad (1.4)$$

Where ∂Q is the capacity of variation of nanoparticles during jet-spray coating and $\partial \theta$ is the angular variation of the jet-spray gun during nanoparticle coating. The change in the capacity of variation of nanoparticles and the variation of the angle of the jet-spray gun impact membrane clusters and must be studied for an optimal operation that decreases the formation of clusters during jet-spray coating. From Bernoulli's equation, we can get the change in pressure during jet-spray coating impacting the speed of nanoparticles at different coating angles that impact the formation of membrane clusters. The volume flow rate between the two points can be given by the continuity equation, $V_2 = (A_1V_1/A_2)$. The change in pressure during jet-spray coating is given as:

$$\Delta P = \frac{1}{2}\rho \left[\frac{A_1}{A_2} V_1 \right]^2 - \frac{1}{2}\rho V_1^2 \qquad (1.5)$$

where ρ is the density of nanoparticles and V_1 the velocity of nanoparticles at entrance A_1 in the jet-spray gun and A_2 is the area of the jet-spray gun at the discharge. The derived model of the function of coating pressure, the coating angle of the jet-spray gun and the distance moved by the spray gun during coating, was tested on the model of nanoparticle coating and scattering of nanoparticles on the ceramic membrane surface and on the model of surface energy driven separability as derived by Sob et al (2020) and is given by Equation (1.6) as [32]:

$$r = r_0 - \frac{2\lambda}{\lambda + n} r_p \qquad (1.6)$$

where r_0 is the size of the aperture without coated nanoparticles, λ the density of nano-particles coated on the membrane channel and n the maximum number of particles that can be coated on the membrane channel surface to give a complete membrane smoothness that leads to lowest surface energy. Equations (1.1–1.6) are solved simultaneously using engineering equation solver software (f-chart software, Madison, w153744, USA.

1.2.2.2 Manufacturing of Ceramic Membrane Surface by Jet-spray

Ceramic hydrophobic nanoparticles, ceramics, Nano4Stone nanoparticles and a spray gun were purchased for the experiment. The samples were crushed into small grain sizes of 8 mm^2 as shown in Figure 1.6(a). Ceramic material was washed with a pre-clean detergent two times to remove foreign particles, which may result in preventing proper blending of nanoparticles on the membrane during coating as shown in Figure 1.6(b). Ceramic material was then allowed to dry under room temperature for 24 hours as shown in Figure 1.6(c).

Before coating, the uncoated ceramic membrane samples were put separately in a zip-lock plastic bag to be sent for microscopic analysis as shown in Figure 1.6(g). Jet-spray coating was done under different coating rounds. During coating, the jet-spray gun used for coating was kept 24 mm away from the membrane surface at an angle of 9° with reference from the vertical axis to the membrane surface as shown in Figure 1.6(d). These parameters were maintained when coating all ceramic beads. Four coating rounds were employed on different ceramic grains for both LP and HP. Membranes were coated in the following order: 1st coat, 2nd coat, 3rd coat and

4th coat for both HP and LP rounds. The coating was done in two-minute intervals between coating rounds for both LP and HP. After coating, the ceramic grains were dried up for 24 hours at room temperature as shown in Figure 1.6(e). The coated ceramic membrane samples were then put separately in zip-lock plastic bags to be sent for microscopic analysis as shown in Figure 1.6(f). Nano4Stone sample was put separately in a small plastic jar to be sent for microscopic analysis to observe membrane cluster minimization due to varying nanoparticles scattering under varying coating pressure that gave different orientations, morphology, spatial distribution, and sizes of nanoparticles on the membrane surface as shown in Figure 1.6(h).

After the membrane material was manufactured using both LP and HP coating rounds, there was a need to send these samples for microscopy analysis. Microscopic analysis is able to analyse the random phenomena of nanoparticle shapes and sizes, the spatial distribution of nanoparticles and morphology of nanoparticles as stated by, hence it was imperative to observe how membrane clusters were minimized during different coating rounds.

1.3 RESULTS AND DISCUSSION

1.3.1 RESULTS AND DISCUSSION OF THE THEORETICAL MODELLING AND SIMULATION OF SURFACE ENERGY-DRIVEN SEPARABILITY

The proposed models derived in this study were tested with the following data from Sob et al (2019), $\rho = 1000$ kg/m^3, h = 6.626×10^{-34} J.s, $\mu = 0.000720$ m^2/s, $S_1 = 0.3$ m, Vvol = 0.12 m^3, $t_2 = 150$ sec, $t_3 = 120$ sec, $A_1 = 0.08$ m, $A_2 = 0.04$ m, F = 100 kN. P = 1000, $S_1 = 0.3$, V = 200 m/s, $t_2 = 3$ sec, $t_3 = 1$ sec, $\sigma = 0.002$, $A_1 = 0.08$ m, $A_2 = 0.0$ 4 m, F = 100 kN. The obtained results are presented and discussed.

The obtained results shown in Figure 1.7(a-b) revealed the relationship between total force during jet-spray coating and jet spray distance and cluster sizes during the coating process. It was revealed as shown in Figure 1.7(a-b) that increasing the total force in the jet-spray gun and coating distances increases the sizes of membrane clusters during jet-spray coating as shown in Figure 1.7(b). This is because the mobility of nanoparticles vibrates at a higher speed during the coating process and this impacts the scattering of nanoparticles which creates more membrane clusters during the coating process. It was also shown that as the coating distance increases with the coating total force in the jet-spray gun, there was an optimal coating distance that gave optimal membrane cluster minimization during the coating process as shown in Figure 1.7(a-b). The cluster size minimization also gave optimal surface spread of nanoparticles in the membrane surface which gave a smooth membrane surface and a lower surface energy to increase membrane wettability. This can be explained since as the coating process began, more membrane clusters were created, which created a rough membrane surface. As more coating processes take place, a rough membrane surface becomes a smooth membrane surface due to the fact that there increase in the formation of membrane clusters that

(a)

Jet Spray Distance During Coating [MM]

(b)

Cluster Sizes During Coating (NM)

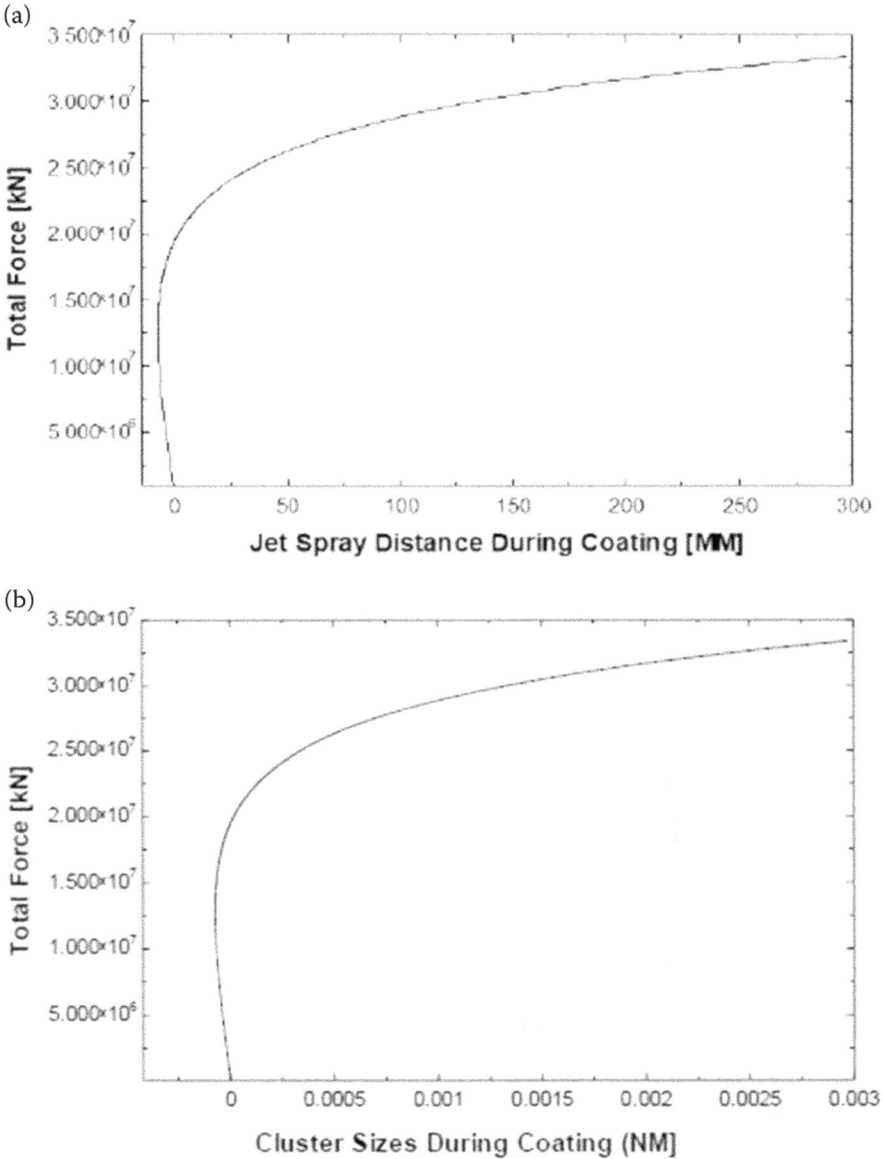

FIGURE 1.7   (a) Jet spray distance during coating [mm] against Total Force from the jet-spray gun [kN] (b) Cluster sizes during coating [nm] against Total Force from the jet-spray gun during coating [kN].

were stabilized, which lead to the creation of smooth membrane surfaces which lowered surface energy and increased membrane wettability. The relationship between jet spray angle during coating and total force during coating and their impact on the change in cluster size is revealed in Figure 1.8(a-b).

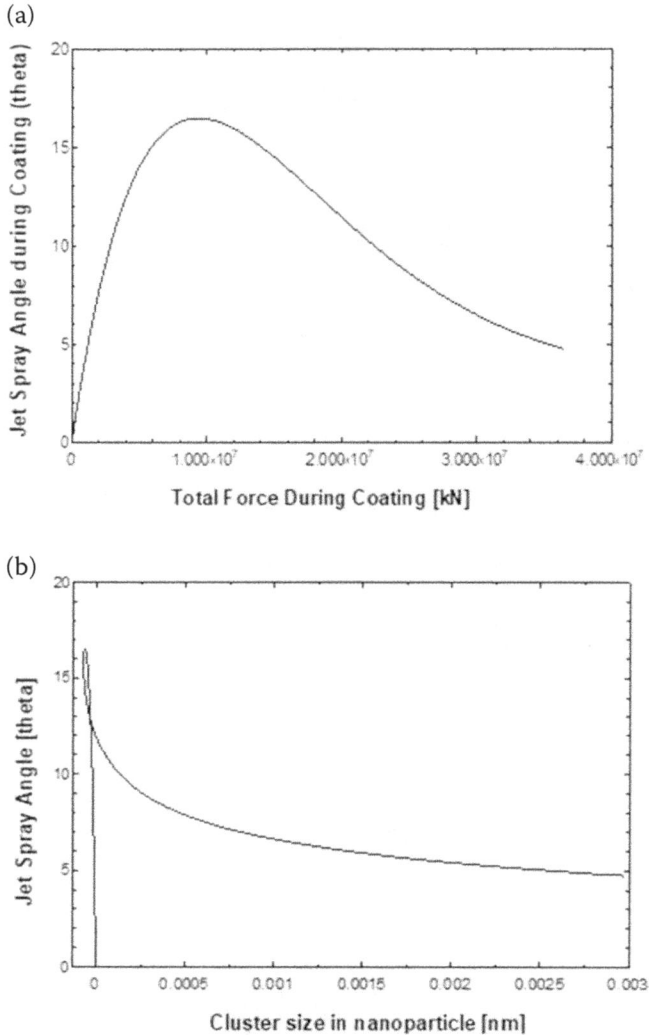

FIGURE 1.8 (a) Jet spray angle during coating [theta] against Total Force from the jet spray [kN] (b) Jet spray angle during coating [theta] against Cluster sizes during coating [nm].

The results in Figure 1.8(a) revealed an increase in jet spray angle and total force in the jet spray during coating to an optimal level which was accompanied by a decrease during the coating process. The increase in coating angles and total forces to an optimal level shown in Figure 1.8(a), lead to a decrease in cluster sizes during the nanoparticle coating process as shown in Figure 1.8(b). It was revealed as shown in Figure 1.8(b) that during the initial process of jet-spray coating more membrane clusters were created, which revealed the initial increase which was accompanied by a continuous decrease at the optimal point during the coating process. The reason for the initial increase which was accompanied by a decrease was that during the initial process of coating more clusters were created on the membrane surface. As more coating takes place, these clusters were minimized due to continuous coating which took place since the required coating forces were produced to minimize the surface cluster on the membrane surface. Therefore, at optimal coating forces and at optimal coating angles, membrane clusters were minimized leading to smoother membrane surfaces which lowered surface energy and improved surface wettability. The formation of membrane clusters during the initial process of coating created a more rough membrane surface which increased surface energy. As the coating pressure and coating angles increased these clusters were minimized since the rough surfaces became smooth surfaces which lowered surface energy and improved membrane surface wettability as shown in Figure 1.9(a-b).

Figure 1.9(a) revealed that membrane surface energy increases gradually until optimum levels of surface energy are reached on a rough surface. This was because when rough membrane surfaces were initially generated during the initial process of jet-spray coating, the membrane surface energy immediately increased which decreased surface wettability. At the optimal point of the coating when the membrane surface clusters began to minimize, membrane rough surfaces became smooth surfaces which lowered surface energy and increased membrane wettability during oil-water separation. The change in membrane clusters also impacts the aperture sizes. This is due to the degree of surface roughness to smoothness during cluster minimization leads to a decrease in aperture sizes and this lowers surface energy and improves surface wettability.

1.3.2 Results and Discussion of the Manufactured Ceramic Membrane with Minimized Clusters on the Surface for Efficient Wettability during Oil-water Separation

It was imperative to analyse all the samples to observe the surface properties. The samples were sent for microscopic observation for the characterization of nanoparticles. Scanning Electron microscopy (SEM) and Energy Dispersion Spectroscopy (EDS) were opted for observation on the samples sent for microscopic analysis. SEM ensures accuracy and validation during characterization and EDS reveals the degree of elements on the sample and the variation of intensity and kilo electrons volt (keV) on a full scale [8,9,32].

1.3.2.1 Ceramic Control Sample and Nanoparticles Control Sample

Figure 1.10(a) shows the uncoated ceramic sample with a rough membrane surface and Figure 1.10(b) shows EDS for the uncoated ceramic sample depicting oxygen with the highest peak or intensity, followed by silicon, iron, aluminium, potassium, magnesium and sodium. Figure 1.10(c) depicts the Nano4stone control sample and Figure 1.10(d) EDS for the Nano4stone control sample depicting only fluorine (F) as the control element of hydrophobic nanoparticles, which created membrane hydrophobicity for the oil/water separation process. Therefore, F was the main

(a)

(b)

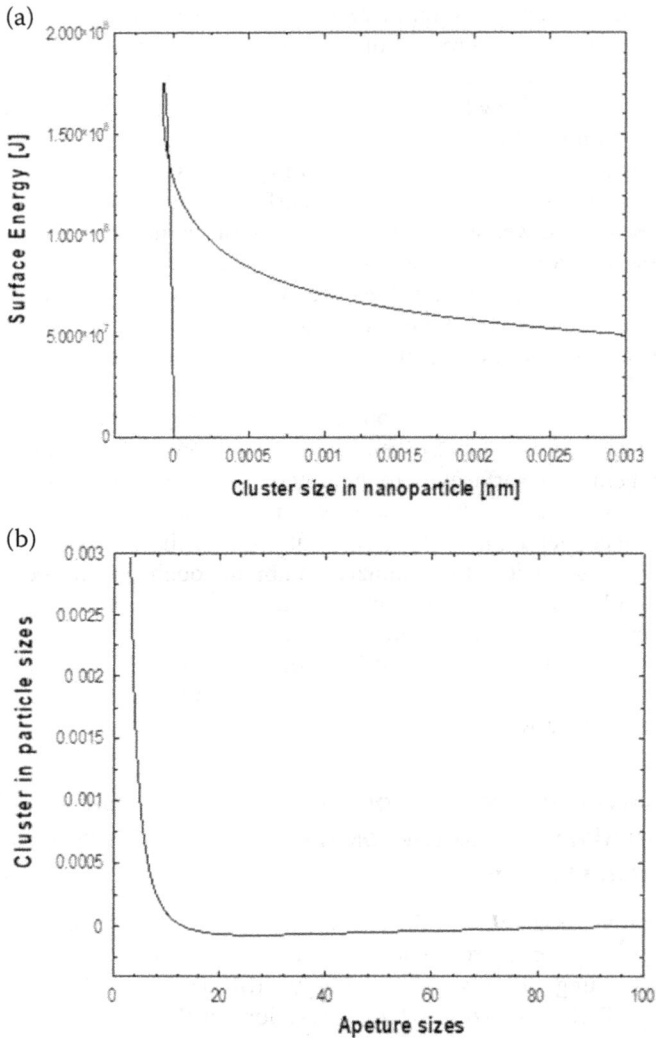

FIGURE 1.9 (a) Surface Energy [J] against Cluster sizes in nanoparticles during coating [nm] (b) Cluster particle sizes during coating [nm] against aperture sizes during coating.

scattering element which varied during HP and LP coating rounds. Since both control samples had been tested and analysed, it was necessary to examine the rounds of HP and LP coating.

1.3.2.2   LP Coating Round One and Round Two

Figure 1.11(a) depicts the 1st LP coating sample and Figure 1.11(b) shows EDS for the 1st LP coating sample showing a very high content of oxygen, followed by silicon, aluminium, iron, potassium, fluorine, magnesium and sodium. Clusters were

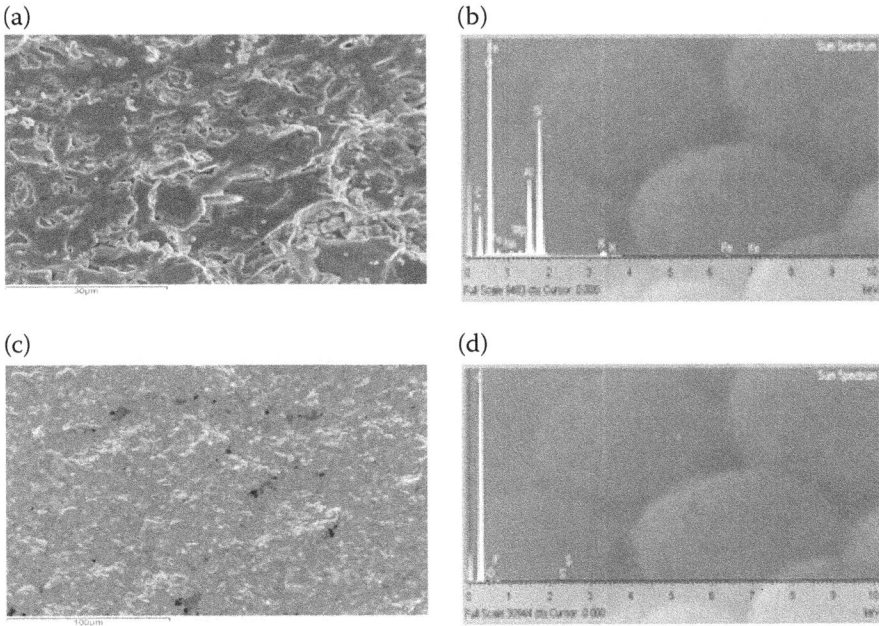

FIGURE 1.10   (a) Ceramic control sample, (b) EDS, (c) Hydrophobic nanoparticles control sample and (d) EDS.

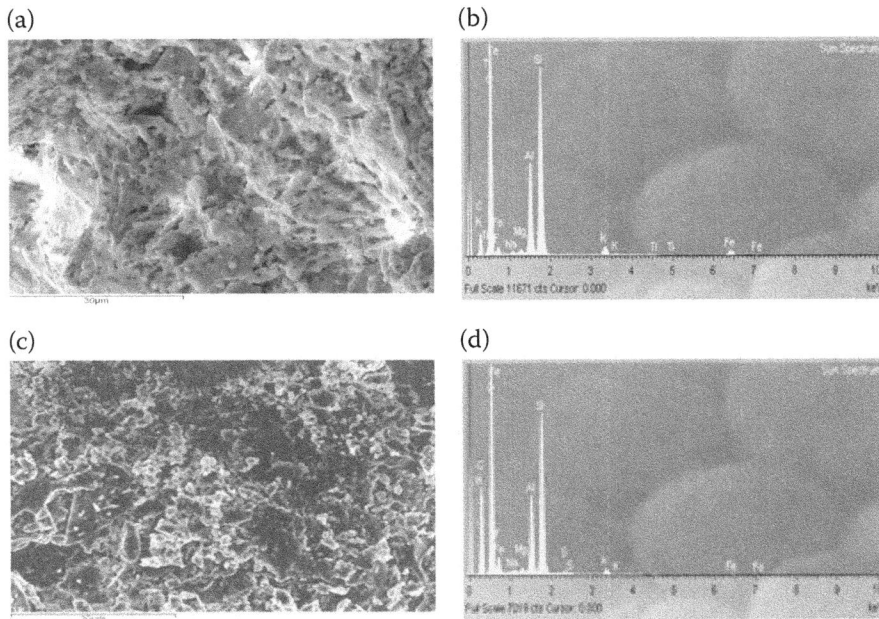

FIGURE 1.11   (a) 1st LP coating ceramic membrane, (b) 1st LP coating (EDS), (c) 2nd LP coating ceramic membrane and (d) 2nd LP coating (EDS).

observed as shown in Figure 1.11(a). The cluster sizes are bigger, resulting in an inhomogeneous surface. This indicates a rougher surface on the membrane, and this doesn't improve membrane wettability on the surface. Figure 1.11(c) depicts the 2nd LP coating sample and Figure 1.11(d) shows EDS for the 2nd LP coating sample showing a very high content of oxygen, followed by silicon, aluminium, potassium, iron, magnesium and sodium. A more visible distribution of clusters was observed as shown in Figure 1.11(c). This is due to that the coating force, coating distance and coating angle and were varied between 0.2×10^7 kN & 2.4×10^7 kN, 10 mm & 24 mm and 1° & 9° respectively in this coating round in trying to identify the optimal levels of coating for cluster minimization. It was therefore imperative to study the 3rd and 4th rounds of LP coating.

1.3.2.3   LP Coating Round Three and Round Four

Figure 1.12(a) depicts the 3rd LP coating sample and Figure 1.12(b) shows EDS for the 3rd LP coating sample showing a very high content of oxygen, followed by silicon, potassium, aluminium, iron and magnesium while sodium. Small clusters were observed as shown in Figure 1.12(a) when compared with the 1st LP coating and 2nd LP coating. There's still a visible distribution of clusters. The coating force, coating distance and coating angle were increased from 0.2×10^7 kN to 2.4×10^7 kN, 10 mm to 24 mm and 1° to 9° respectively in trying to identify the optimal levels of coating for cluster minimization. This authenticates why clusters are minimized better on the

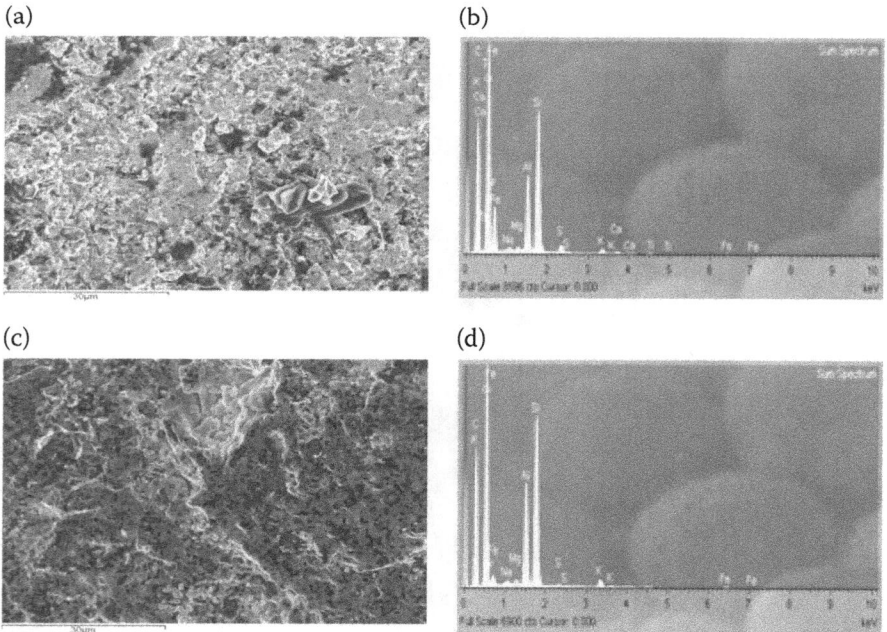

(a)   (b)

(c)   (d)

FIGURE 1.12   (a) 3rd LP coating ceramic membrane, (b) 3rd LP coating (EDS), (c) 4th LP coating ceramic membrane and (d) 4th LP coating (EDS).

3rd LP coating round as compared to the 1st LP coating and 2nd LP coating. Figure 1.12(c) depicts the 4th LP coating sample and Figure 1.12(d) shows EDS for the 4th LP coating sample showing a very high content of oxygen, followed by silicon, potassium, aluminium, iron, magnesium and sodium. The coating distance and coating angle were decreased from 2.4×10^7 to 0.2×10^7 kN, 24 mm to 10 mm and $9°$ to $1°$ in trying to get the optimal levels of coating to get more clusters minimized. The coating levels maintained in this coating round didn't reach the optimal level therefore it gave poor membrane wettability as compared to the 3rd LP coating. This, therefore, produced poor surface wettability on the membrane surface. It was therefore important to study HP coating and its impact on wettability.

1.3.2.4 HP Coating Round One and Round Two

Figure 1.13(a) depicts the 1st HP coating sample and Figure 1.13(b) shows EDS for the 1st HP coating sample showing a very high content of oxygen, followed by silicon, aluminium, iron, potassium, fluorine, sodium, magnesium and sulphur. Few clusters were observed when compared with the 1st LP, 2nd LP, 3rd LP and 4th LP coating, as shown in Figure 1.13(a). Clusters are observed on the surface. This resulted in a rough membrane surface produced and this impacted membrane wettability negatively. The coating force was increased from 0.2×10^7 kN to 2.4×10^7 kN in this coating round to get the optimal level of coating pressure. This membrane gave better membrane wettability as compared to a 1st LP, 2nd LP, 3rd LP and 4th LP coating but still,

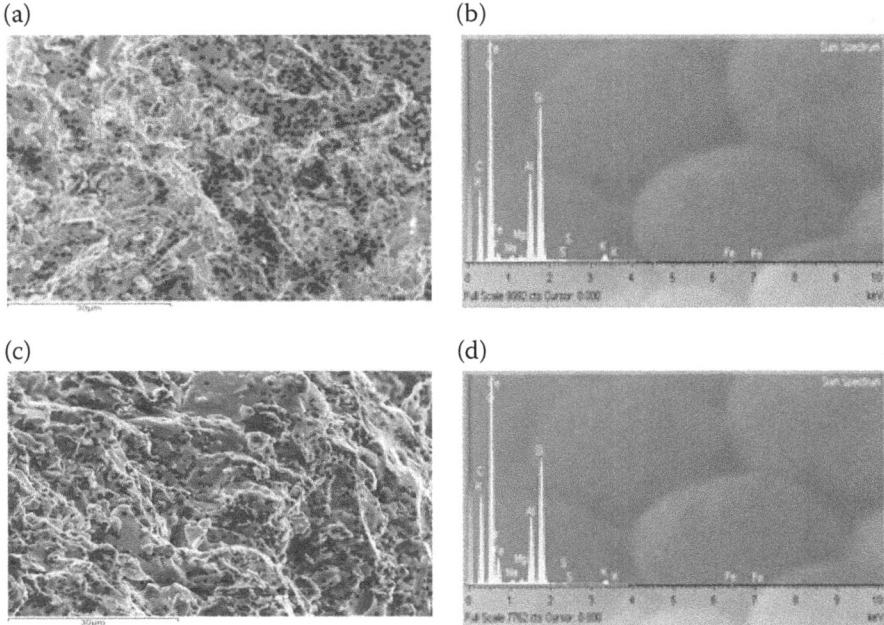

FIGURE 1.13 (a) 1st HP coating ceramic membrane (b) 1st HP coating (EDS), (c) 2nd HP coating ceramic membrane and (d) 2nd HP coating (EDS).

clusters were observed on the membrane, resulting in an inhomogeneous membrane. Figure 1.13(c) depicts the 2nd HP coating sample and Figure 1.13(d) shows EDS for the 2nd HP coating sample showing a very high content of oxygen, followed by silicon, aluminium, fluorine, iron, potassium, magnesium, sodium and sulphur. Few clusters were observed when compared with the 1st LP, 2nd LP, 3rd LP, 4th LP and 1st HP coating. This still created a rough membrane surface. It was therefore imperative to study the 3rd and 4th rounds of HP coating to achieve a membrane with better wettability for oil-water separation.

1.3.2.5 HP Coating Round Three and Round Four

Figure 1.14(a) depicts the 3rd HP coating sample and Figure 1.14(b) shows EDS for the 3rd HP coating showing a very high content of Oxygen, followed by Silicon, Aluminium, Iron, Fluorine, Sulphur, Potassium, Sodium and Magnesium. Fewer but visible clusters were observed when compared with 1st LP, 2nd LP, 3rd LP, 4th LP, 1st HP and 2nd HP coating. The surface was still inhomogeneous due to visible clusters observed on the surface. It, therefore, was imperative to do 4th round of HP coating to produce a membrane surface with further reduced clusters to give a smoother surface with more minimized clusters gives better surface wettability. Figure 1.14(c) depicts the 4th HP coating sample and Figure 1.14(d) shows EDS for the 4th HP coating showing a very high content of Oxygen, followed by Silicon, Fluorine, Aluminium, Iron, Sulphur, Potassium, Sodium and Magnesium. This membrane gave lesser clusters observed on the surface when compared with 1st LP,

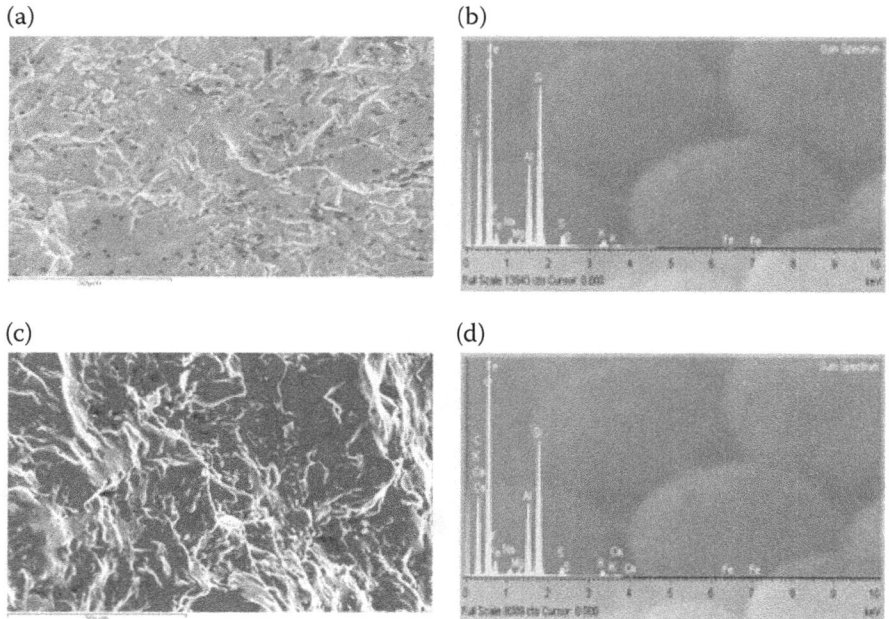

FIGURE 1.14   (a) 3rd HP coating ceramic membrane (b) 3rd HP coating (EDS), (c) 4th HP coating ceramic membrane and (d) 4th HP coating (EDS)

2nd LP, 3rd LP, 4th LP, 1st HP, 2nd HP and 3rd HP coating. The scattering of F is even on the membrane surface due to surface uniformity, thus creating a smoother membrane surface. The produced membrane during this coating is more homogeneous as compared to 1st LP, 2nd LP, 3rd LP, 4th LP, 1st HP, 2nd HP and 3rd HP coating. This was achieved by increasing the coating force, coating distance and coating angle to 2.4×10^7 kN, 24 mm and 9°. The pressure reached during this coating round reached an optimal level to produce a smoother membrane. This improved membrane wettability on the surface when related to the lotus effect on surface wettability. Since the produced membrane after 4st HP coating offered a smoother membrane surface as compared to all other coating rounds, there was no need to study 5th round HP coating to see if the produced membrane surface will be more efficient with stable wettability.

REFERENCES

[1] N. Gao and Z. K. Xu, "Ceramic membranes with mussel-inspired and nanostructured coatings for water-in-oil emulsions separation," *Sep. Purif. Technol.*, vol. 212, no. November 2018, pp. 737–746, 2019, doi: 10.1016/j.seppur.2018.11.084.

[2] L. Chen, K. Guan, W. Zhu, C. Peng, and J. Wu, "Preparation and mechanism analysis of high performance ceramic membrane by spray coating," pp. 39884–39892, 2018, doi: 10.1039/c8ra07258b.

[3] S. Rezaei, H. Abadi, M. Reza, M. Hemati, F. Rekabdar, and T. Mohammadi, "Ceramic membrane performance in micro fi ltration of oily wastewater," *DES*, vol. 265, no. 1–3, pp. 222–228, 2011, doi: 10.1016/j.desal.2010.07.055.

[4] H. Wang *et al.*, "Review: Porous Metal Filters and Membranes for Oil–Water Separation," *Nanoscale Res. Lett.*, vol. 13, 2018, doi: 10.1186/s11671-018-2693-0.

[5] Z. Chu, Y. Feng, and S. Seeger, "Oil/Water Separation with Selective Superantiwetting/ Superwetting Surface Materials Angewandte," pp. 2328–2338, 2015, doi: 10.1002/ anie.201405785.

[6] J. S. Weston, R. E. Jentoft, B. P. Grady, D. E. Resasco, and J. H. Harwell, "Silica Nanoparticle Wettability: Characterization and E ff ects on the Emulsion Properties," 2015, doi: 10.1021/ie504311p.

[7] M. Abbasi, M. Mirfendereski, M. Nikbakht, M. Golshenas, and T. Mohammadi, "Performance study of mullite and mullite – alumina ceramic MF membranes for oily wastewaters treatment," *DES*, vol. 259, no. 1–3, pp. 169–178, 2010, doi: 10.1016/ j.desal.2010.04.013.

[8] P. B. Sob, A. A. Alugongo, and T. B. Tengen, "Scanning Electron Microscopy, Energy Dispersive X-ray Spectroscopy and Statistica Analysis of High and Low Pressure Coatings on Sediments Membrane for Stable and Efficient Wettability," vol. 12, no. 12, pp. 1–22, 2019.

[9] P. B. Sob, A. A. Alugongo, and T. B. Tengen, "The Stochastic Effect of Nanoparticles Inter-Separation Distance on Membrane Wettability During Oil/Water Separation," vol. 13, no. 5, pp. 842–866, 2020.

[10] T. Chen, M. Duan, and S. Fang, "Fabrication of novel superhydrophilic and underwater superoleophobic hierarchically structured ceramic membrane and its separation performance of oily wastewater," *Ceram. Int.*, vol. 42, no. 7, pp. 8604–8612, 2016, doi: 10.1016/j.ceramint.2016.02.090.

[11] Y. Wang, X. Wang, Y. Liu, S. Ou, Y. Tan, and S. Tang, "Re fi ning of biodiesel by ceramic membrane separation," vol. 90, pp. 422–427, 2009, doi: 10.1016/j.fuproc. 2008.11.004.

[12] M. Ben, N. Hamdi, M. A. Rodriguez, and K. Mahmoudi, "Preparation and characterization of new ceramic membranes for ultra fi ltration," vol. 44, no. September 2017, pp. 2328–2335, 2018, doi: 10.1016/j.ceramint.2017.10.199.

[13] F. Zhang, W. Bin Zhang, Z. Shi, D. Wang, J. Jin, and L. Jiang, "Nanowire-haired inorganic membranes with superhydrophilicity and underwater ultralow adhesive superoleophobicity for high-efficiency oil/water separation," *Adv. Mater.*, vol. 25, no. 30, pp. 4192–4198, 2013, doi: 10.1002/adma.201301480.

[14] A. Ditsch, P. E. Laibinis, D. I. C. Wang, and T. A. Hatton, "Controlled Clustering and Enhanced Stability of Polymer-Coated Magnetic Nanoparticles," no. 13, pp. 6006–6018, 2005.

[15] X. Zheng, Z. Guo, D. Tian, X. Zhang, W. Li, and L. Jiang, "Underwater self-cleaning scaly fabric membrane for oily water separation," *ACS Appl. Mater. Interfaces*, vol. 7, no. 7, pp. 4336–4343, 2015, doi: 10.1021/am508814g.

[16] K. J. Kubiak, M. C. T. Wilson, T. G. Mathia, and P. Carval, "Wettability versus roughness of engineering surfaces," *Wear*, vol. 271, no. 3–4, pp. 523–528, 2011, doi: 10.1016/j.wear.2010.03.029.

[17] M. Shahien and M. Suzuki, "Low power consumption suspension plasma spray system for ceramic coating deposition," *Surf. Coatings Technol.*, vol. 318, pp. 11–17, 2017, doi: 10.1016/j.surfcoat.2016.07.040.

[18] Y. Cai, T. W. Coyle, G. Azimi, and J. Mostaghimi, "Superhydrophobic Ceramic Coatings by Solution Precursor Plasma Spray," *Nat. Publ. Gr.*, pp. 1–7, 2016, doi: 10.1038/srep24670.

[19] P. Fauchais and A. Vardelle, "Innovative and emerging processes in plasma spraying: from micro- to nano-structured coatings," 2011, doi: 10.1088/0022-372 7/44/19/194011.

[20] R. Balzarotti, C. Cristiani, and L. F. Francis, "Combined dip-coating/spin-coating depositions on ceramic honeycomb monoliths for structured catalysts preparation," *Catal. Today*, vol. 334, no. June 2018, pp. 90–95, 2019, doi: 10.1016/j.cattod. 2019.01.037.

[21] L. Hao, Z. Chen, R. Wang, C. Guo, P. Zhang, and S. Pang, "A non-aqueous electrodeposition process for fabrication of superhydrophobic surface with hierarchical micro/nano structure," *Appl. Surf. Sci.*, vol. 258, no. 22, pp. 8970–8973, 2012, doi: 10.1016/j.apsusc.2012.05.130.

[22] L. H. Chen *et al.*, "Nanostructure depositions on alumina hollow fiber membranes for enhanced wetting resistance during membrane distillation," *J. Memb. Sci.*, vol. 564, pp. 227–236, 2018, doi: 10.1016/j.memsci.2018.07.011.

[23] T. Van Gestel, D. Sebold, W. A. Meulenberg, M. Bram, and H. Buchkremer, "Manufacturing of new nano-structured ceramic – metallic composite microporous membranes consisting of ZrO 2, Al2O3, TiO 2 and stainless steel ☆," vol. 179, pp. 1360–1366, 2008, doi: 10.1016/j.ssi.2008.02.046.

[24] R. Ferrando, J. Jellinek, and R. L. Johnston, "Nanoalloys: From Theory to Applications of Alloy Clusters and Nanoparticles," vol. 108, no. 3, 2008.

[25] Y. Cai *et al.*, "A smart membrane with antifouling capability and switchable oil wettability for high-efficiency oil/water emulsions separation," *J. Memb. Sci.*, vol. 555, no. March, pp. 69–77, 2018, doi: 10.1016/j.memsci.2018.03.042.

[26] T. Meng *et al.*, "Nano-structure construction of porous membranes by depositing nanoparticles for enhanced surface wettability," *J. Memb. Sci.*, vol. 427, pp. 63–72, 2013, doi: 10.1016/j.memsci.2012.09.051.

[27] L. Zhang, J. Wu, Y. Wang, Y. Long, N. Zhao, and J. Xu, "Combination of Bioinspiration: A General Route to Superhydrophobic Particles," 2012.

[28] J. E. Zhou, Q. Chang, Y. Wang, J. Wang, and G. Meng, "Separation of stable oil-water emulsion by the hydrophilic nano-sized ZrO2 modified Al2O3 microfiltration

membrane," *Sep. Purif. Technol.*, vol. 75, no. 3, pp. 243–248, 2010, doi: 10.1016/ j.seppur.2010.08.008.

[29] M. Miwa, A. Nakajima, A. Fujishima, K. Hashimoto, and T. Watanabe, "Effects of the surface roughness on sliding angles of water droplets on superhydrophobic surfaces," *Langmuir*, vol. 16, no. 13, pp. 5754–5760, 2000, doi: 10.1021/la991660o.

[30] D. Quéré, "Wetting and roughness," *Annu. Rev. Mater. Res.*, vol. 38, pp. 71–99, 2008, doi: 10.1146/annurev.matsci.38.060407.132434.

[31] G. Hurwitz, G. R. Guillen, and E. M. V. Hoek, "Probing polyamide membrane surface charge, zeta potential, wettability, and hydrophilicity with contact angle measurements," *J. Memb. Sci.*, vol. 349, no. 1–2, pp. 349–357, 2010, doi: 10.1016/ j.memsci.2009.11.063.

[32] P. B. Sob, A. A. Alugongo, and T. B. Tengen, *"Controllability and stability of selectively wettable nanostructured membrane for oil / water separation,"* A thesis submitted in fulfilment of the requirements for the degree Doctorate Technologiae in Mechanical Engineering in the Faculty of Engineering & Technology, 2020

[33] N. Gao, Y. Y. Yan, X. Y. Chen, and D. J. Mee, "Superhydrophobic surfaces with hierarchical structure," *Mater. Lett.*, vol. 65, no. 19–20, pp. 2902–2905, 2011, doi: 10.1016/j.matlet.2011.06.088.

[34] T. Ogi, F. Iskandar, Y. Itoh, and K. Okuyama, "Characterization of dip-coated ITO films derived from nanoparticles synthesized by low-pressure spray pyrolysis," pp. 343–350, 2006, doi: 10.1007/s11051-005-9006-0.

2 Fabrication of Biopolymer Composite Film Derived from Agro-waste Possessing Antioxidant Properties

Malvika Sharma, Preeti Beniwal, and Amrit Pal Toor

CONTENTS

2.1 INTRODUCTION

Food packaging is vital for the food packaging industry as it protects the food from contamination and maintains the quality of the product. Presently, petroleum-based plastics like polyethylene and polystyrene are being used due to their low cost,

DOI: 10.1201/9781003242291-2

durability and good mechanical properties [1,2]. However, this plastic packaging has a very slow degradation rate leading to negative impacts on the environment like reducing soil fertility, causing air pollution during incineration, along with polluting the oceans, etc. Plastic waste thus has triggered the development of biodegradable plastics synthesized from the biopolymers like cellulose, starch, chitosan, etc., which are biodegradable, non-toxic and recyclable [3,4]. Among these, cellulose is one of the widely utilized biopolymers for the synthesis of bio-degradable composites due to its abundance in nature, along with non-toxicity, renewability, eco-friendly nature and low cost. Agricultural wastes like rice straw, wheat straw, and sugarcane bagasse have cellulose content of around 40 w/w%, 32.9 w/w%, 30.2 w/w%, respectively, hence these agro wastes hold great potential for the formulations of various composite [5,6].

Films made from Cellulose Fibre (CF) are thermally stable and strong with high tensile strength [5–7]. Thus, CF has shown huge potential in the packaging industry. However, the inflexibility of the cellulose-based films has been one of the limiting factors due to the strong hydrogen bonding between cellulose polymer chains [8]. This inflexibility can be overcome by the addition of a plasticizer. Plasticizers reduce hydrogen bonding between cellulose polymeric chains and enhance hydrogen bonding between cellulose polymer and plasticizer. Plasticizers also have the capability to improve the mechanical strength and elasticity of the films [8,9]. A lot of plasticizers like glycerol, sorbitol, polyethylene glycol, etc. are available which have been incorporated into the films and proved to amplify the elasticity in films [5,10,11]. Among all these, the utilization of polyethylene glycol (PEG) proved to be prominent due to its non-toxicity, biocompatible nature, non-antigenic, supports biodegradation rate of cellulose and lower hydrophilicity than glycerol [12–15]. Moreover, it has tremendous compatibility with cellulose nanofibers (CNF) due to strong hydrogen bonding constructing biodegradable and biocompatible composites with CNF [16–18].

Recently, research is oriented towards the development of active packaging systems, which are designed to contain components that increase the shelf life or preserve the condition of the packed food product. Active packages with antioxidant properties have gained a lot of attention because they're one of the most promising alternatives to standard packaging as they minimize food oxidation, which is one of the leading causes of food spoilage [19]. Various organic acids like citric acid, succinic acid, tartaric acid, adipic acid and malic acid are widely distributed in nature and are known to possess antioxidant properties due to the presence of hydroxyl groups [20]. Citric acid (CA) is a six-carbon tricarboxylic acid $6(C_6H_8O_7.H_2O)$, a common metabolite of plants and animals and is abundantly available in citrus fruit juice whereas tartaric acid (TA) is a natural dicarboxylic acid existing as pair of enantiomers and an achiral meso compound and is naturally present in grapes and cranberries [21,22]. Both of these organic acids are a good cross-linker and are Generally Recognized as Safe (GRAS) [23]. The inclusion of CA/TA has improved the tensile strength of starch/poly (butylene adipate co-terephthalate) blown films [24]. Furthermore, both of these organic acids depicted antioxidant characteristics, and it has been found that formulation films with organic acids resulted in the increased shelf life of certain food items [20,25,26].

The current study proposes the extraction of CF from rice straw and the synthesis of CF film loaded with PEG denoted as CF/PEG. Furthermore, TA and CA incorporated CF/PEG film denoted as CF/PEG/TA and CF/PEG/CA were formed. The study was framed for the evaluation of the antioxidant capacity of the TA and CA-incorporated CF/PEG films. These organic acid-induced films could be used in the food industries for food packaging due to their inherent antioxidant properties.

2.2 MATERIALS AND METHODS

2.2.1 RAW MATERIAL AND CHEMICALS

Rice straws were collected from the agricultural fields of Anandpur Sahib, Punjab, India and were chemically treated to procure cellulose fibres. Polyethylene glycol (Mn 400) purchased from Sigma-Aldrich, citric and tartaric acid purchased from Qualigens, sodium hydroxide procured from E. Merck India Limited, HCl purchased from Chemigens Research and Fine Chemicals, were utilized in film preparation. Following chemicals viz hydrogen peroxide, ferrous sulphate heptahydrate and 1,10-Phenanthroline AR purchased from Central Drug House Pvt. Ltd. (New Delhi, India), sodium dihydrogen phosphate monohydrate, di sodium hydrogen phosphate dihydrate and calcium chloride purchased from Merck Life Sciences Pvt. Ltd. (Mumbai, India), were utilized to analyse the antioxidant activity of films.

2.2.2 ISOLATION OF CELLULOSE FIBRES FROM RICE STRAW

The CF was extracted using a chemical treatment method as previously reported but with a few tweaks [27]. Rice straw fibres were cut into small pieces of 4–5 mm, washed with tap water and dried for 1–2 days in order to eradicate the dirt and foreign materials. These clean rice straw pieces were then treated in a Soxhlet apparatus with 2:1, v/v toluene/ethanol mixture for up to four cycles to remove waxes, oils and pigments. These dried dewaxed fibres were then steeped in 8% NaOH at 50°C for 4 h to remove the present hemicellulose and lignin. The fibres were bleached by soaking them in 8% hydrogen peroxide at 60°C for 2 h to get rid of any remnant lignin. These dried delignified fibres were then hydrolysed with 5M HCl at 50°C for 12 h breaking the hydrogen bonds between cellulose polymer, separating them and removing the residual components like pectin followed by washing with running water till pH 7 was acquired and were air dried till constant weight is achieved. The acquired fibres were then pulverized in a grinder and were further employed for film fabrication. The isolated fibres are depicted in Figure 2.1.

2.2.3 SYNTHESIS OF BIO COMPOSITE FILM

The composite films were synthesized via solvent casting method [7]. The film-forming solutions were prepared by mixing CF, PEG, TA/CA. PEG served as a plasticizer. The TA and CA content varied from 10wt% to 40wt%. CF was dissolved by adding 1wt% CF in a beaker and was continuously stirred at 250 rpm for 30 min using a magnetic stirrer. This solution was then sonicated in an ultra-bath sonicator for 60 min. To this solution, 30% w/w PEG and TA/CA were added and

FIGURE 2.1 Synthesized films: (a): extracted CF; (b) CF/PEG film; (c) CF/PEG/30wt %TAfilm(optimized); (d) CF/PEG/20wt%CA film (optimized).

further mixed for 1 h on a hot plate. The major advantage of using TA and CA organic acids is that they act as cross-linkers as well as antioxidants. The same technique was used to synthesize neat CF/PEG film, but no other ingredients were added to the mixture. The solutions were then poured into Petri dishes with a diameter of 90 mm and kept for drying at room temperature for 48 h. The films so formed were then removed from the Petri dishes and further utilized for various physiochemical analyses.

2.2.4 CHARACTERIZATION OF SYNTHESIZED FILMS

A scanning electron microscope (JSM-6010LA, JEOL Co. USA) was used to investigate the morphology of the synthesized films. FTIR spectra were recorded using the Perkin Elmer-Spectrum RX-IFTIR instrument ranging from 40 to 4000 cm^{-1}. It was performed in order to analyse the functional groups and to provide insights into the structural characteristics of the films. X-Ray diffraction (PAN Analytical Xpert Pro) was used to analyse the crystalline nature of the synthesized films over the 2θ range of $0°$ and $90°$. The thermal properties of the synthesized films were determined by TGA analyser (Perkin Elmer STA-6000) with a heating rate of 10°C /min within the temperature range of 25°C to 500°C. The contact angle (Kruss advance 1.6.2.0) was performed using the sessile drop method for the estimation of the wetting properties of the solid surface. The mechanical properties of the synthesized films were investigated using Universal Testing Machine (Biss, India). The water vapour permeability of the samples was conducted based on the ASTM method E96-80(American Society for Testing and Materials, 1987) [28] and the radical scavenging activity of the film was measured using Hydroxyl radical scavenging assay [29].

2.3 RESULTS AND DISCUSSION

2.3.1 MORPHOLOGY OF THE SYNTHESIZED FILMS

The scanning electron micrographs of extracted CF, CF/PEG, CF/PEG/30wt%TA (optimum) and CF/PEG/20wt%CA (optimum) film are shown in Figure 2.2. The CF as shown in Figure 2.2(a) appeared to be rod or ribbon-like in structure, had a rough surface and were present in clusters. Acid hydrolysis led to the destruction of hydrogen bonds present between cellulose polymers resulting in CF separation [7]. Figure 2.2(b) shows the rough surface of the film with the addition of PEG to the CF. A continuous network without holes denoted strong hydrogen bonding between CF and PEG. Low molecular weight PEG consists of huge accessible hydroxyl groups and smaller molecular volume which allow diffusion and interaction of PEG with polymer matrix [5]. Figure 2.2(c) & (d) represents CF/PEG films with 30wt%

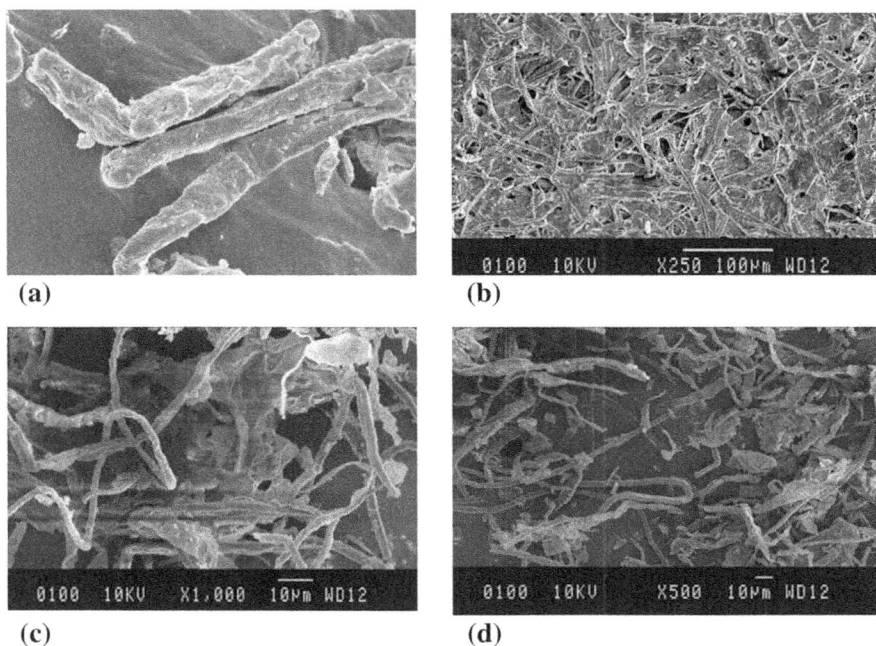

(a)   (b)

(c)   (d)

FIGURE 2.2 SEM micrographs of: (a) CF (b) CF/PEG film (c) CF/PEG/30wt%TA film (d) CF/PEG/20wt%CA film.

TA and 20wt% CA, respectively. With the addition of these organic acids to the CF/PEG blend, fold structures with interpenetrating holes were observed which are the result of cross-linking between hydroxyl groups on CF, PEG and carboxyl groups on TA and CA [30]. These groups are further confirmed with FTIR technique and the results are mentioned later.

2.3.2 FOURIER TRANSFORM INFRARED SPECTROSCOPY (FTIR) STUDY OF THE SYNTHESIZED FILMS

The FTIR spectra of the CF and synthesized films are shown in Figure 2.3. The absorption band of CF at 3413 cm^{-1} corresponded to OH stretching vibration as well as to the intermolecular and intramolecular hydrogen bonds among cellulose polymers [31] and the one at 2919 cm^{-1} was correlated with CH stretching. In the CF/PEG film, the absorption band at 3342 cm^{-1} was due to OH stretching vibration and the one at 2870 cm^{-1} was due to CH stretching. This shifting of bands on the incorporation of PEG into CF confirmed the presence of hydrogen bond interaction between CF and PEG [32]. Another absorption band in the range 1300–1050 cm^{-1} represented C-O-C stretching of glycosidic linkages between CF and PEG [16].

In the FTIR-spectra of composite films, the absorption bands corresponding to OH stretching were registered at 3341 cm^{-1} and 3339cm^{-1} belonging to CF/PEG/30wt%TA (optimum) and CF/PEG/20wt%CA (optimum) respectively. This decrease in the OH intensity peak could be due to chemical interaction with organic acids [31]. The absorption band at 2919.5 cm^{-1} and 2920.7 cm^{-1} for CF/PEG30wt%TA and CF/PEG 20wt%CA respectively represented CH stretching and in the range 1300–1050 cm^{-1} corresponded to C-O-C stretching of the glycosidic linkages between CF, PEG and the organic acids. A significant absorption band in the range

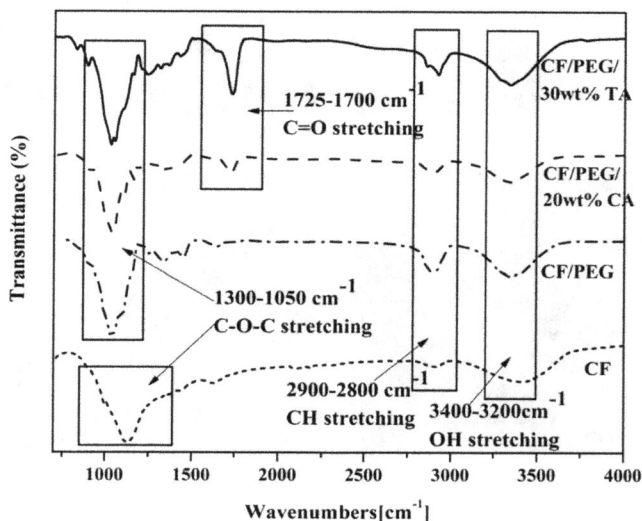

FIGURE 2.3 FTIR spectra of CF and synthesized films.

$1750-1730$ cm^{-1} was observed representing ester bond formation, confirming the interaction between CF/PEG and the organic acids [26]. Thus, hydrogen bond formation between CF, PEG and TA/CA was confirmed from the above results.

2.3.3 X-Ray Diffraction of the Synthesized Films

Figure 2.4 illustrates XRD diffractograms for the CF/PEG, CF/PEG/30wt%TA and CF/PEG/20wt%CA films. On incorporation of PEG into CF, the characteristic peaks observed were at 15.9° and 22.3°. Slight broadening and reduction of the peaks at 22.3° were due to PEG loading on the amorphous fraction of CF [33]. The CF/PEG film with 30wt% TA had characteristic peaks at 16.2° and 22.6° and the CF/PEG film with 20wt%CA had characteristic peaks at 16.0° and 22.5°. However, some of the crystalline peaks of CF almost disappeared indicating that the crystalline structure of CF was disrupted by the addition of PEG, and organic acids TA and CA due to reinforcement between the matrices [16].

2.3.4 Contact Angle

The contact angle of the films on the addition of PEG to CF was observed to be 0° which could be due to the availability of free hydroxyl groups on the film with the addition of PEG. On addition of TA and CA in the CF/PEG film, a reduction in the hydrophilicity of the films was observed [34] (See Table 2.1). The addition of TA to the CF/PEG combination, increased the contact angle to a maximum of 113.6° with respect to CF/PEG film, leading to a hydrophobic surface while the addition of CA to the CF/PEG combination resulted in a hydrophilic surface. However, it had an improved contact angle than CF/PEG film. High contact angle indicates low surface energy resulting in low polar components. The free hydroxyl groups on CF and

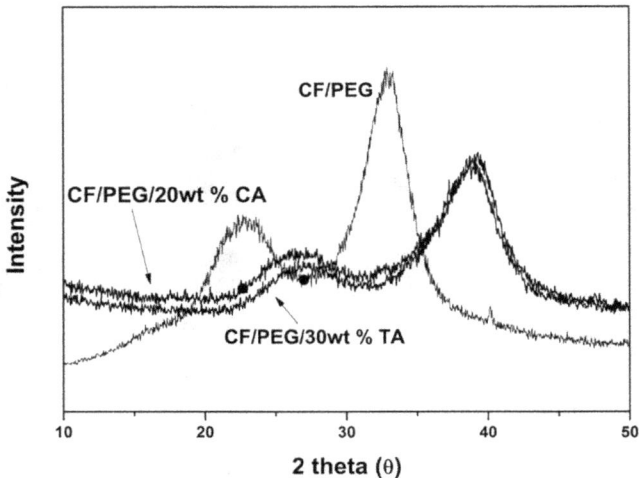

FIGURE 2.4   X-Ray diffractograms of synthesized films.

TABLE 2.1

Contact angle of CF/PEG films with various amounts (wt %) of TA and CA

CP film with TA (wt %)	Contact Angle	CP film with CA (wt %)	Contact Angle
10	123.2°	10	88.7°
20	121.1°	20	88.9°
30	117.6°	30	0
40	113.6°	40	0

PEG got cross-linked with carboxyl groups on organic acids as shown by FTIR analysis, resulting in hydrophobic characteristics of the film's surface as compared to CF/PEG film.

2.3.5 ANTIOXIDANT ACTIVITY

The generation of free radicals on oxidation leads to a huge loss of food products. It causes foul smell generation, bad colour, bad flavour etc making the food unsuitable for consumption [26]. Recently, the focus has been on the development of an active food packaging system for the preservation of food products. The addition of antioxidants to the food packaging material is one of the strategies to render activity against oxidizing agents, hence preserving packed food. The antioxidant activity was determined using a hydroxyl radical scavenging assay [29]. The brief procedure follows as.

The antioxidant activity of all the films with TA and CA were observed and are depicted in Figure 2.5. The observed antioxidant activity for 10wt%, 20wt%, 30wt% and 40wt% TA were 13.51%, 15.46%, 16.04% and 16.22%, respectively and the observed antioxidant activity for 10wt%, 20wt%, 30wt% and 40wt% CA were

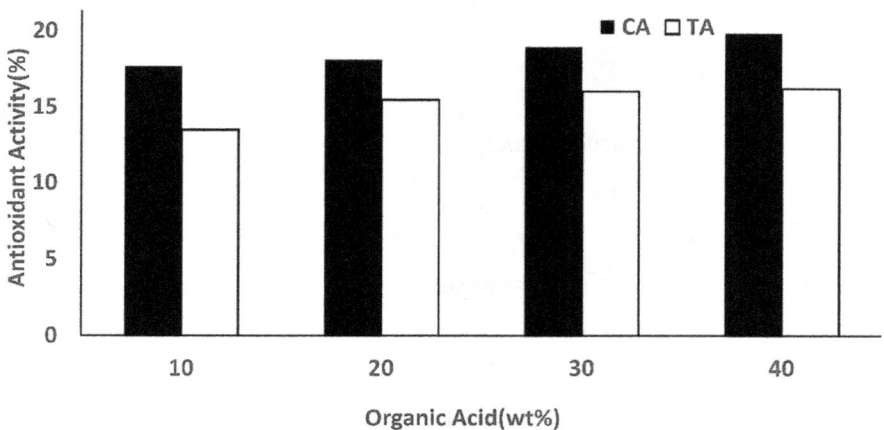

FIGURE 2.5 Antioxidant activity of films synthesized with organic acids.

TABLE 2.2

Antioxidant activity of CF/PEG film containing various amounts (wt %) of TA and CA

Amount of organic acids (wt%) in films	Antioxidant activity of films with CA (%)	Antioxidant activity of films with TA (%)
10	17.68	13.51
20	18.13	15.46
30	18.97	16.04
40	19.86	16.22

17.68%, 18.13%, 18.97% and 19.86%, respectively. The activity increased with an increasing amount of TA/CA [35]. All the films with CA had higher antioxidant activity than all the films containing TA indicating CA to be a better antioxidant agent than TA as depicted in Table 2.2. The CF/PEG film with 30wt% TA/20wt%CA showed good contact angle and antioxidant activity. Thus, we performed further tests with CF/PEG film, CF/PEG/30wt%TA film and CF/PEG/20wt%CA film.

2.3.6 WATER VAPOUR PERMEABILITY

Table 2.3 depicts the observed water vapour permeability (WVP) values of all the synthesized films. The addition of 30wt%TA and 20wt%CA to the CF/PEG film increased the WVP. Both the films with TA and CA had higher WVP than the CF/PEG film. This increment in the WVP could be due to the presence of free hydroxyl groups which take part in hydrogen bonding with the water vapours and also low molecular weight PEG relaxes the CF chains leading to easy diffusion of water vapours through the synthesized films [5]. The WVP of fish gelatin film has also been increased with increasing amounts of CA [36].

2.3.7 THERMAL ANALYSIS

Thermal stability of the films was tested using a thermogravimetric analyser (TGA) and the TGA and DTG curves are shown in Figure 2.6(a) and (b) respectively. For

TABLE 2.3

Water vapour permeability of the synthesized films

Type of film	Water vapor permeability $\times 10^{-5}$ (g/mhPa)
CF/PEG	7.87
CF/PEG/30wt% TA	8.45
CF/PEG/20 wt%CA	8.47

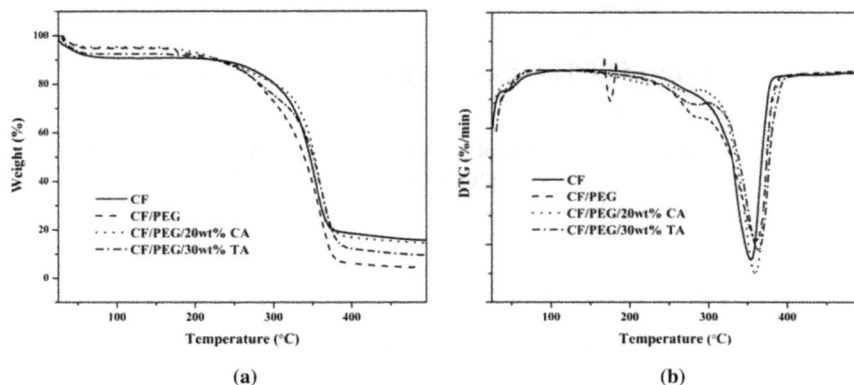

(a)   (b)

FIGURE 2.6   (a) TGA curve of CF and synthesized films; (b) DTG curve of CF and synthesized films.

CF the first stage of mass reduction was observed below 100°C corresponding to evaporation of water. The second stage of mass reduction occurred in the temperature range of 250°C to 380°C which indicated degradation of CF. Similar results have been observed where cellulose degradation occurred in the temperature range of 200°C to 400°C [37]. On addition of PEG to CF the first stage of weight loss was observed below 100°C attributing to water evaporation. The second stage of mass reduction was observed in the temperature range 250°C to 390°C which corresponded to degradation of PEG [16], CF. Further incorporation of 30wt%TA to CF/PEG blend the first stage of mass reduction was observed below 100°C corresponding to water evaporation and the second stage of weight loss was observed in the temperature range 300°C to 396°C representing degradation of CF and PEG. Incorporation of 20wt% CA to CF/PEG blend also had the first stage of mass reduction below 100°C attributed to water evaporation. The second stage of mass reduction was observed between 290°C and 395°C corresponding to degradation of CF and PEG. The percent weight loss of the synthesized films on Thermo gravimetric analysis (Table 2.4) represented higher weight loss of the synthesized films than CF. On incorporation of 30wt%TA and 20wt%CA to the CF/PEG blend a shift in the temperature range in the second stage has been observed indicating their

TABLE 2.4

Weight loss percent of synthesized films

Synthesized films	Wt$_L$ %
CF	78.55
CF/PEG	91.2
CF/PEG/30wt%TA	86.9
CF/PEG/20wt%CA	80.8

thermal stability which is due to inter-intramolecular hydrogen bonding [20]. On observing DTG curves it could be analysed that the addition of 30wt%TA/20wt %CA to CF/PEG blend improved the thermal stability of the synthesized films [1].

2.3.8 MECHANICAL PROPERTIES

A food packaging material must have good strength and flexibility in order to be able to package a food product. In a composite film, it is very necessary to attain interfacial properties of the components in order to get good mechanical properties as poor dispersion might lead to a weak film. Figure 2.7(a) shows the tensile strength of the synthesized films before permanent deformation. It was observed that the film containing CF and PEG had 0.09 MPa tensile strength due to intermolecular hydrogen bonding between CF and PEG. On addition of TA and CA to the CF/PEG film, the tensile strength observed was 0.25 MPa and 0.59 MPa, respectively. Both the films with TA and CA had higher tensile strength than the film with CF/PEG only. The increment in the tensile strength of the films with these organic acids was due to intermolecular hydrogen bonding between CF, PEG and organic acids. An improvement in the tensile strength of the polyvinyl alcohol/ carboxymethyl cellulose film has also been reported with the addition of CA which acted as a cross-linker resulting in toughening of the internal bonds of the biopolymers [38].

Figure 2.7(b) depicts Elongation at break (%) (EAB) which is related to the stretchability of the films before it breaks. It was observed that the film containing only CF and PEG had EAB 24.2 %. With the addition of TA to the CF/PEG film the EAB increased to 31.9% and when CA was added the EAB increased to 85.1%. Both

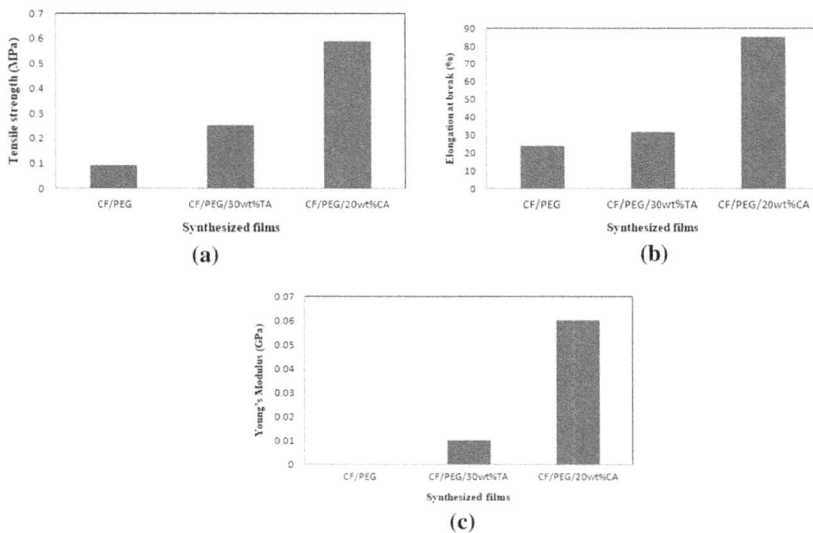

FIGURE 2.7   (a) Tensile strength of the synthesized films; (b) Elongation at break of the synthesized films; (c) Young's Modulus of the synthesized films.

TABLE 2.5

Tensile strength (MPa), elongation at break (%), Young's Modulus (GPa) values of the CF/PEG films with the addition of TA and CA

Film composition	Tensile strength (MPa)	Elongation at break (%)	Young's Modulus (GPa)
CF/PEG	0.09	24.2	0
CF/PEG/30wt%TA	0.25	31.9	0.01
CF/PEG/20wt%CA	0.59	85.1	0.06

the films with TA and CA had higher EAB than the film with no organic acid. Both the organic acids improved the EAB of the film. The hydroxyl groups of PEG form hydrogen bonding with cellulose polymers and carbonyl groups of TA/CA leading to increased free volume and decreased intermolecular forces and glass transition temperature causing an increase in the flexibility [1] as depicted in Table 2.5.

Figure 2.7(c) shows Young's modulus (GPa) which is related to the rigidity of films. Higher Young's modulus indicates rigid material. There was no Young's modulus observed for the film with only CF and PEG as it had very low tensile strength. On the addition of TA to the CF/PEG film, a slight improvement in Young's modulus of 0.01 GPa was observed. Furthermore, on addition of CA to CF/PEG film Young's modulus increased to 0.06 GPa. Both the films with CA and TA had a higher Young's modulus than the CF/PEG film. An increase in Young's modulus of the starch/poly (butylene adipate co-terephthalate) has also been observed on addition of TA [39].

2.4 CONCLUSION

Bio composite films with various amounts of TA and CA were effectively fabricated. The incorporation of TA and CA to the CF/PEG combination modified the physical and chemical properties of the CF/PEG-based film bringing significant improvement.

The film incorporated with TA had a hydrophobic surface whereas CA-containing film had a hydrophilic surface. Both the films showed antioxidant activity making them suitable for an active packaging system. The films with 30wt% TA/20wt% CA showed good contact angle and antioxidant activity. So, further tests of the films were performed with these amounts of TA and CA.

Both the films with organic acids had increased water vapour permeability which could be due to the presence of PEG, relaxing the CF chains and leading to easy diffusion of water vapours through the film and can be improved by the addition of some lipids in future research. The thermal stability and mechanical properties were also improved with the addition of 30wt%TA and 20wt%CA to the CF/PEG combination which might be due to strong cross-linking between the components of the film. To summarise, 30wt%TA and 20wt% CA are the optimum amount for adding to the CF/PEG blend and also have the potential for active food packaging due to good antioxidant activity as well as good mechanical properties.

ACKNOWLEDGEMENT

All the authors acknowledge TEQIP-III (MHRD, Govt. of India), Dr. SSB University Institute of Chemical Engineering and Technology, Panjab University, Chandigarh for their financial support.Declaration of Competing Interest: The authors declare that they have no known competing financial interests or personal relationships that could have appeared to influence the work reported in this paper.

REFERENCES

[1] Z. Wu, J. Wu, T. Peng, Y. Li, D. Lin, B. Xing, C. Li, Y. Yang, L. Yang, L. Zhang, and R. Ma, "Preparation and application of starch/polyvinyl alcohol/citric acid ternary blend antimicrobial functional food packaging films," *Polymers (Basel)*, vol. 9, no. 3, pp. 102, 2017.

[2] S. Shankar, X. Teng, G. Li, and J.-W. Rhim, "Preparation, characterization, and antimicrobial activity of gelatin/ZnO nanocomposite films," *Food Hydrocoll.*, vol. 45, pp. 264–271, 2015.

[3] V. Siracusa, P. Rocculi, S. Romani, and M. D. Rosa, "Biodegradable polymers for food packaging: a review," *Trends Food Sci. Technol.*, vol. 19, no. 12, pp. 634–643, 2008.

[4] Y. Freile-Pelegrín, T. Madera-Santana, D. Robledo, L. Veleva, P. Quintana, and J. A. Azamar, "Degradation of agar films in a humid tropical climate: Thermal, mechanical, morphological and structural changes," *Polym. Degrad. Stab.*, vol. 92, no. 2, pp. 244–252, 2007.

[5] R. H. F. Faradilla, G. Lee, P. Sivakumar, M. Stenzel, and J. Arcot, "Effect of polyethylene glycol (PEG) molecular weight and nanofillers on the properties of banana pseudostem nanocellulose films," *Carbohydr. Polym.*, vol. 205, pp. 330–339, 2019.

[6] P. K. Sadh, S. Duhan, and J. S. Duhan, "Agro-industrial wastes and their utilization using solid state fermentation: a review," *Bioresour. Bioprocess.*, vol. 5, no. 1, 2018.

[7] H. Abral, J. Ariska, M. Mahardika, D. Handayani, I. Aminah, N. Sandrawati, A. B. Pratama, N. Fajri, S. M. Sapuan, and R. A. Ilyas, "Transparent and antimicrobial cellulose film from ginger nanofiber," *Food Hydrocoll.*, vol. 98, pp. 105266, 2020.

[8] P. Parameswara, T. Demappa, R. T. Guru, and R. Somashekar, "Microstructural parameters of hydroxypropylmethylcellulose films using X-ray data," *Iranian Polymer Journal*, vol. 17, no. 11, pp. 821–829, 2008.

[9] C. Maraveas, "Production of sustainable and biodegradable polymers from agricultural waste," *Polymers (Basel)*, vol. 12, no. 5, pp. 1127, 2020.

[10] Tian, D. Liu, Y. Yao, S. Ma, X. Zhang, and A. Xiang, "Effect of sorbitol plasticizer on the structure and properties of melt processed polyvinyl alcohol films: Polyvinyl alcohol films … ," *J. Food Sci.*, vol. 82, no. 12, pp. 2926–2932, 2017.

[11] S. Mali, M. V. E. Grossmann, M. A. García, M. N. Martino, and N. E. Zaritzky, "Effects of controlled storage on thermal, mechanical and barrier properties of plasticized films from different starch sources," *J. Food Eng.*, vol. 75, no. 4, pp. 453–460, 2006.

[12] P. Kunthadong, R. Molloy, P. Worajittiphon, T. Leejarkpai, N. Kaabbuathong, and W. Punyodom, "Biodegradable plasticized blends of poly (L-lactide) and cellulose acetate butyrate: from blend preparation to biodegradability in real composting conditions," *Journal of Polymers and the Environment*, vol. 23, no. 1, pp. 107–113, 2015.

[13] T. Behjat, A. R. Russly, C. A. Luqman, A. Y. Yus, and I. Nor Azowa, "Effect of PEG on the biodegradability studies of Kenaf cellulose-polyethylene composites," *International Food Research Journal*, vol. 16, no. 2, pp. 243–247, 2009.

[14] N. A. Alcantar, E. S. Aydil, and J. N. Israelachvili, "Polyethylene glycol-coated biocompatible surfaces," *J. Biomed. Mater. Res.*, vol. 51, no. 3, pp. 343–351, 2000.

[15] Khairuddin, E. Pramono, S. B. Utomo, V. Wulandari, A. Zahrotul W, and F. Clegg, "The effect of polyethylene glycol Mw 400 and 600 on stability of Shellac Waxfree," *J. Phys. Conf. Ser.*, vol. 776, p. 012054, 2016.

[16] S. Gopi, A. Amalraj, N. Kalarikkal, J. Zhang, S. Thomas, and Q. Guo, "Preparation and characterization of nanocomposite films based on gum arabic, maltodextrin and polyethylene glycol reinforced with turmeric nanofiber isolated from turmeric spent," *Mater. Sci. Eng. C Mater. Biol. Appl.*, vol. 97, pp. 723–729, 2019.

[17] M. Gu, C. Jiang, D. Liu, N. Prempeh, and I. I. Smalyukh, "Cellulose nanocrystal/ poly (ethylene glycol) composite as an iridescent coating on polymer substrates: Structure-color and interface adhesion," *ACS applied materials & interfaces*, vol. 8, no. 47, pp. 32565–32573, 2016.

[18] M. A. Ghalia and Y., Dahman, "Fabrication and enhanced mechanical properties of porous PLA/PEG copolymer reinforced with bacterial cellulose nanofibers for soft tissue engineering applications," *Polym*, vol. 61, pp. 114–131, 2017.

[19] C. López-de-Dicastillo, J. Gómez-Estaca, R. Catalá, R. Gavara, and P. Hernández-Muñoz, "Active antioxidant packaging films: Development and effect on lipid stability of brined sardines," *Food Chem.*, vol. 131, no. 4, pp. 1376–1384, 2012.

[20] W. Zhang, Q. Jiang, J. Shen, P. Gao, D. Yu, Y. Xu and W. Xia, "The role of organic acid structures in changes of physicochemical and antioxidant properties of cross-linked chitosan films," *Food Packaging and Shelf Life.*, vol. 31, p. 100792, 2022.

[21] A. R. Angumeenal and D. Venkappayya, "An overview of citric acid production," *Lebenson. Wiss. Technol.*, vol. 50, no. 2, pp. 367–370, 2013.

[22] M. Stratford, "PRESERVATIVES | traditional preservatives – organic acids," in *Encyclopedia of Food Microbiology*, Elsevier, 1999, pp. 1729–1737.

[23] E. Mani-López, H. S. García, and A. López-Malo, "Organic acids as antimicrobials to control Salmonella in meat and poultry products," *Food Res. Int.*, vol. 45, no. 2, pp. 713–721, 2012.

[24] J. B. Olivato, M. V. E. Grossmann, A. P. Bilck, and F. Yamashita, "Effect of organic acids as additives on the performance of thermoplastic starch/polyester blown films," *Carbohydr. Polym.*, vol. 90, no. 1, pp. 159–164, 2012.

[25] R. Hraš, M. Hadolin, Ž. Knez, and D. Bauman, "Comparison of antioxidative and synergistic effects of rosemary extract with α-tocopherol, ascorbyl palmitate and citric acid in sunflower oil," *Food Chem.*, vol. 71, no. 2, pp. 229–233, 2000.

[26] R. Priyadarshi, Sauraj, B. Kumar, and Y. S. Negi, "Chitosan film incorporated with citric acid and glycerol as an active packaging material for extension of green chilli shelf life," *Carbohydr. Polym.*, vol. 195, pp. 329–338, 2018.

[27] M. Dilamian and B. Noroozi, "A combined homogenization-high intensity ultra-sonication process for individualizaion of cellulose micro-nano fibers from rice straw," *Cellulose*, vol. 26, no. 10, pp. 5831–5849, 2019.

[28] ASTM E96-80, "Standard test methods for water vapour transmission of material", Annual book of ASTM, Philadelphia, PA: American Society for Testing and Materials, 1987.

[29] G.-R. Zhao, H. M. Zhang, T. X. Ye, Z. J. Xiang, Y. J. Yuan, Z. X. Guo, and L. B. Zhao, "Characterization of the radical scavenging and antioxidant activities of danshensu and salvianolic acid B," *Food Chem. Toxicol.*, vol. 46, no. 1, pp. 73–81, 2008.

[30] Z. Zhang, Y. Li, L. Song, L. Ren, X. Xu, and S. Lu, "Swelling resistance and water-induced shape memory performances of sisal cellulose nanofibers/polyethylene glycol/citric acid nanocellulose papers," *J. Nanomater.*, vol. 2019, pp. 1–9, 2019.

[31] M. G. Raucci, M. A. Alvarez-Perez, C. Demitri, D. Giugliano, V. De Benedictis, A. Sannino, and L. Ambrosio, "Effect of citric acid crosslinking cellulose-based

hydrogels on osteogenic differentiation: Effect of Cellulose-Based Hydrogels on Osteogenic Differentiation," *J. Biomed. Mater. Res. A*, vol. 103, no. 6, pp. 2045–2056, 2015.

[32] D. Mishra, P. Khare, D. K. Singh, S. Luqman, P. A. Kumar, A. Yadav, T. Das, and B. K. Saikia, "Retention of antibacterial and antioxidant properties of lemongrass oil loaded on cellulose nanofibre-poly ethylene glycol composite," *Ind. Crops Prod.*, vol. 114, pp. 68–80, 2018.

[33] Butchosa Robles, N. (2014). Tailoring cellulose nanofibrils for advanced materials (Doctoral dissertation, KTH RoyalInstitute of Technology).

[34] M. Lim, H. Kwon, D. Kim, J. Seo, H. Han, and S. B. Khan, "Highly-enhanced water resistant and oxygen barrier properties of cross-linked poly (vinyl alcohol) hybrid films for packaging applications," *Progress in Organic Coatings*, vol. 85, pp. 68–75, 2015.

[35] A. Ounkaew, P. Kasemsiri, K. Kamwilaisak, K. Saengprachatanarug, W. Mongkolthanaruk, M. Souvanh, U. Pongsa, and P. Chindaprasirt, "Polyvinyl alcohol (PVA)/starch bioactive packaging film enriched with antioxidants from spent coffeeground and citric acid," *J. Polym. Environ.*, vol. 26, no. 9, pp. 3762–3772, 2018.

[36] J. Uranga, A. I. Puertas, A. Etxabide, M. T. Dueñas, P. Guerrero, and K. de la Caba, "Citric acid-incorporated fish gelatin/chitosan composite films," *Food Hydrocoll.*, vol. 86, pp. 95–103, 2019.

[37] F. M. Pelissari, P. J. do A. Sobral, and F. C. Menegalli, "Isolation and characterization of cellulose nanofibers from banana peels," *Cellulose*, vol. 21, no. 1, pp. 417–432, 2014.

[38] S. R. Kanatt and S. H. Makwana, "Development of active, water-resistant carboxymethyl cellulose-poly vinyl alcohol-Aloe vera packaging film," *Carbohydr. Polym.*, vol. 227, no. 115303, p. 115303, 2020.

[39] J. B. Olivato, M. M. Nobrega, C. M. O. Müller, M. A. Shirai, F. Yamashita, and M. V. E. Grossmann, "Mixture design applied for the study of the tartaric acid effect on starch/polyester films," *Carbohydr. Polym.*, vol. 92, no. 2, pp. 1705–1710, 2013.

3 Design, Development and Analysis of an Improved Tricycle

Partha Pratim Dutta, Polash P. Dutta, and Mrinmoy G. Baruah
Tezpur University, Assam, India

CONTENTS

DOI: 10.1201/9781003242291-3

3.1   INTRODUCTION

Considering the present Indian scenario of the low-income group like rickshaw pullers, the tricycle (rickshaw) is a very cheap means of short-distance transportation in both city and rural areas. The rickshaw is propelled by human energy to carry an average of two full-grown persons. Therefore, these manually driven rickshaws provide the livelihood of eight million very poor rickshaw pullers' families. In this regard, a study on the existing design of the conventional model of tricycle rickshaw reveals that traditional rickshaws use old-age technology and poor mechanical design, and therefore they are non-ergonomic in maneuverability. The origin and present status of the tricycle rickshaw may be defined as follows [1–3].

3.1.1   TRICYCLE RICKSHAW FIRST DEVELOPMENT

The historical root of the tricycle rickshaw was linked to definite local transportation means. The first tricycle appeared after 1750 AD. It was 150 years before the invention of the original bicycle with two wheels and no motor power. It was about two hundred years ahead of mass-manufactured chain and sprocket bicycles. German watchmaker Stephan Farffler devised a primitive tricycle. He was a person with a disability and fabricated a small tricycle with three wheels. The tricycle had a simple mechanism with respect to the front wheel. It facilitated the rotation of the crank for transferring power to the driven wheel in the tricycle. Two French inventors devised the first pedal-operated tricycle in 1797 and named this mechanism a tricycle. This name is still a synonym with this local manually driven transportation unit from that time and onward. A tricycle is sometimes named a trike, and it is a manually driven three-wheeled unit. Some tricycles, like cycle rickshaws for local transportation of passenger and freight trikes, are used for the commercial requirement in the developing world, like Asia and Africa. Another study reported an estimation of rolling resistances of the tricycle. In the west, adult tricycle sizes are mostly used for recreation, shopping, and fitness (battery-powered and solar-powered). Children and older adults equally like tricycles because they seem to have stability compared to a bike. However typical trikes have poor dynamic lateral stability, and the user must be careful to keep the trike from tipping. Unconventional designs like recumbents have a lower center of gravity and need less attention [1–5].

3.1.2 HYBRID TRICYCLE RICKSHAW

Hybrid tricycles are propelled by another prime mover like an electric motor in addition to the human power. The human and electric drive may be in parallel or series configuration depending upon load-carrying capacity and reliability. A study was reported on a hybrid tricycle with a 400 W 24 V DC motor drive, in addition to human power. Both the options worked independently as well as in hybrid mode. The complete setup was designed and fabricated. A human electric hybrid three-wheeler may be utilized as an alternative to local transportation for personal, recreational, and small-scale commercial uses. The three-wheeler was fabricated using locally available materials from the market that assured the clear possibility for its future commercialization and production in the economically viable way [6]. Another similar study reported on the green manufacturing of a hybrid vehicle. Hybrid vehicles with an internal combustion engine propelled by biodiesel-blended fuel and batteries charged by electric power and regenerative-braking have been reported. A lightweight vehicle with low rolling resistance and aerodynamic drag is expected to have the better fuel economy. Some distinct techniques of weight reduction and the use of some lightweight materials were proposed. To minimize cost, energy, and raw material consumption; recycling of used materials was proposed wherever possible [7]. A permanent magnet motor (PMDC) was used for the hybrid tricycle. Since no internal combustion engine was used, hence pollution created by its operation was almost zero. Therefore, the propulsion of this hybrid tricycle protects the ecosystem and environment from ever-increasing greenhouse gas emissions. Moreover, the proposed design used dual spring shock absorbers that were safe for drivers when the bumping took place in bad road conditions. Additionally, the innovative differential was used so that the power transmits correspondingly to both wheels avoiding the biasing and banking angle (pulling the tricycle on one side). Old and special persons (PWD) may use this vehicle as they get tired in normal conventional cycles [8]. An analysis has been reported within the house-developed tadpole type hybrid electric tricycle. A comparison has been made on two designs (tadpole and delta) of hybrid tricycles. A clearance of 30 cm between the centre of the steering hub and the base of the breastbone (sternum) headrest was as close to the head as possible (2–3) cm. The Backrest angle was (30–40) degrees, and low seat positions were kept to preserve stability. Maximum load applied was 10.78 kN and the factor of safety was found to be 1.4 from the front impact test. For the roll-over test, the maximum load was 7.0 kN and the factor of safety was 2.4 [9]. Paudel and Kreutzmann [10] designed and developed a solar hybrid tricycle. The computed costs for storing power for one full charge as $ 4.17 only. This charge was able to propel the tricycle for 37.14 km. Dutta and Jash [11] studied problems and the energy consumption pattern of the mechanized trike. They observed that specific energy consumption of a mechanized van varied from 59.31 to 151.24 kJ/person/km relying on the number of passengers.

3.1.3 INNOVATIVE DESIGN ON A SIMILAR MECHANISM

Dutta et al. [12–14] designed, developed and tested innovative mechanisms like CVT-based transmission for an electric vehicle, improved manual lawn mower, and low-cost dynamometer for performance evaluation of an electric vehicle. Bhukya and Duvvuri [15] developed a solar electric bicycle with a PMDC motor. The bicycle was capable of travelling at 30 km without pedaling, and regeneration was possible. Keote et al. [16] developed a solar-powered lawn mower. Neuroula, and Sharma [17] designed an electric differential with an individual rear wheel-powered electric vehicle. Pillai et al. [18] designed one all-terrain vehicle gearbox, and Rao et al. [19] designed and developed a smart cart. Resende, and Gonzaga [20] designed and developed the punch planter and observed that Punch planters reduced the power demand in planting for prospective planters. Stress analysis on a bicycle frame was analyzed, and the design was with ANSYS [21].

3.1.4 Tricycle Design

Types of tricycles that commonly used today are

- Delta tricycles – Tricycles whose two main wheels are placed behind the user.
- Tadpole tricycles – Tricycles whose two main wheels are placed in front of the handlebar.
- Recumbent tricycles – Made in delta and tadpole variations; these tricycles have users sitting in a very low and almost lying down position.
- Convertible – Tricycles which can be transformed between delta and tadpole configurations.
- Children's tricycles – Simple delta designs that are made to be very safe in both indoor and outdoor use.
- Manual tricycles – Tricycles powered by human feet or hands.
- Motorized tricycles – Tricycles powered by a combustible or electric engine.

3.1.5 Suspension System

The suspension system of an automobile separates the body from the wheels. All the power from the pedaling shaft goes to three wheels via the transmission system. When the vehicle moves over irregularities on the road, the vehicle feels the shocks on the wheel and the suspension system acts as a screen or filter at this point. The functions of suspension systems are:

- It supports the weight of the vehicle.
- It cushions the bumps and holes in the road.
- It supports the traction between the tires and the road.
- It holds the wheel in alignment.

3.1.6 Components of Suspension System

The main components of the suspension are given below:

- Springs
- Shock Absorber

The springs support the weight of the vehicle, the load on the vehicle, and the road shocks.

The shock absorber helps to control or to dampen the spring action. Without the Shock Absorber, the spring makes the vehicle move up and down every time the vehicle crosses a bump or hole in the road. Shock absorber allows the basic spring action but quickly dampens out unwanted oscillations that follow. It may be electrically or mechanically controlled.

3.1.7 TYPES OF SPRINGS

- Coil Spring.
- Leaf Spring.
- Torsion Spring.
- Air Spring

The above-mentioned springs are briefly described as follows:

3.1.7.1 Coil Spring

This is the most common type of spring and is, in essence, a heavy-duty torsion bar coiled around an axis. Coil springs compress and expand to absorb the motion of the wheel. The Coiled Spring is made up of a round or tapered steel rod. Tapered rod gives the coil a variable spring rate. As the spring is compressed, its resistance to compress further increases. Coil spring is advantageous as it has lightweight, inexpensive, and maintenance-free.

3.1.7.2 Leaf Spring

Leaf springs are classified as either single or multi-leaf. There are more than one metal leafs stacked on top of each other to make this sort of spring, which acts as a single unit. Leaf springs were initially employed in horse-drawn carriages, and they were used in American automobiles up until 1985. They are still in use today on the majority of trucks and heavy-duty vehicles. There are several other names for this type of spring, including Semi-Elliptical and Cart Spring. It is made of spring steel and is about an inch wide and an inch long. There are two tie holes on either end of the arc for attaching it to the vehicle body. The axle is located in the center. Leaf Springs are useful for locating, and to a lesser extent dampening, in addition to springing action. Although the inner leaf friction dampens the suspension's motion, it is poorly regulated, causing friction. So, manufacturers have tried mono leaf suspensions. The main advantages of the leaf springs are that they are simple in construction, and can easily support the chassis of a vehicle. Moreover, they can control axle damping which leads to the main cause of shock in a vehicle. A leaf spring may be attached to the frame directly at both ends or at one end, usually the front, with the other end attached via a shackle, a short swinging arm. The shackle imitates the leaf spring's tendency to elongate when compressed, resulting in softer springiness. The shackle gives the leaf spring some flexibility, preventing it from breaking under heavy loads.

3.1.7.3 Torsion Spring or Torsion Bar

In order to produce Coil Spring-like behavior, Torsion bars employ twisted steel bars. The vehicle's frame is secured with an end of a bar. The other end is connected to a wishbone, which functions as a lever that travels perpendicular to the torsion bar. When the wheel strikes a bump, vertical motion is transferred to the wishbone and subsequently to the torsion bar by the levering action. The Torsion bar then twists around its axis to generate the spring force. It is durable, easy to use, and small in size.

3.1.7.4 Air Spring

Air springs employ the compressive properties of air to absorb wheel vibrations. This idea has been around for over a century and was first used on horse-drawn buggies. Air suspension enhances the height of the ride by load and speed. Air suspension minimizes the tendency for short, bumpy roads and terrains to bounce when the car is empty. Air suspension improves truck and trailer transport ability by offering better control than the suspension in its entirety. One advantage of air suspension is the added comfort afforded by the ability to adjust the air pressure inside the spring, changing the spring rate and hence the ride quality. When dock plates are lacking, aligning loading docks to the deck level is achievable because varied control over air pressure alters the deck or trailer height.

3.1.7.5 Mathematics of Spring Rate

Spring rate is the ratio used to measure how resistant a spring is to being compressed or expanded during the spring deflection. The magnitude of spring force increases as deflection increases according to Hooke's Law. This can be stated as

$$F = -kx \dots \tag{3.1}$$

Where, F = The force exerts by the spring in N (Newton).

k = The Spring rate
x = The displacement from equilibrium position

The spring rate of a coil spring may be calculated by a simple algebraic equation, or it may be measured in a Spring testing machine. The spring constant can be calculated as follows:

$$k = \frac{d^4 G}{8ND^3}. \tag{3.2}$$

Where, d is the wire diameter, G is the spring material's Shear modulus or modulus of rigidity, N is the number of warps, and D is the diameter of the coil.

3.2 OBJECTIVE

It has been observed that the present tricycle rickshaw design tends to pull in one side due to the presence of conventional differential. As a result, an effort is made to develop a mechanically improved high-strength and ergonomic tricycle rickshaw. It will have an improved differential compared to presently using one along with more than two passengers carrying capacity. As a result, it will be a manually powered tricycle only with better mechanical advantage and capacity. The improved designs of the fabricated manually driven tricycle may be utilized in local transportation, recreational and small-scale commercial uses. The three-wheeler is fabricated using locally available standard materials from the market. Based on the above literature review, the objective of the present work is design and development of an improved tricycle rickshaw having low-cost, high capacity, and energy-efficient.

3.3 PRINCIPAL THEORIES

There are two theories involved in this design of tricycle. The theories are as follows:

3.3.1 THEORY 1

The first theory is based on the sprocket being rotated in one direction; it rotates along with the inner ring. But if the sprocket is rotated in the opposite direction, only the outer part rotates, and the inner part does not rotate with the outer part.

3.3.2 THEORY 2

The second theory is based on a chain and sprocket mechanism; when the driving gear is smaller than the driven gear, the torque required to rotate the driving gear is less than that of when the driving gear is bigger than the driven gear.

3.4 METHODOLOGY

For proper functioning and stability of the tricycle, there are many controlling parameters whose estimations are to be made to ensure the practical functioning of the unit. The vehicle is divided into different sub-assemblies such as the frame, transmission, steering, suspension, and brake.

3.4.1 ASSUMPTIONS FOR DESIGN OF IMPROVED TRICYCLE

- Tricycle has minimum dimensions (2.1 m × 1.5 m × 1.6 m).
- The maximum mass of the tricycle is 80 kg.
- Seats should be coupled to the frame through suspension.
- As much as possible, locally available standard materials will be used.
- Maximum human mechanical power available from the rickshaw puller is 300 W.
- Average mechanical power available from the rickshaw puller is 100 W.

The development design method was applied for the overall improvement of conventional tricycle. Initially, the problems faced by rickshaw pullers with the present design were studied. Accordingly, an improved design was conceptualized by applying the assumptions mentioned above. Initially, the total tractive effort required for propelling the improved tricycle rickshaw was estimated as per mathematical modeling in the next section. This followed engineering drawing and design work done using CATIA V5 commercial software. This followed analysis of frame and chain drive using ANSYS commercial software. Finally, manufacturing was initiated as per the flowchart.

3.4.2 TRACTIVE EFFORT REQUIREMENT

The rolling resistance of the improved tricycle is given by Equation (3.3) [4].

$$F_r = C_{rc} W \tag{3.3}$$

Where F_r is rolling resistance (Newton) between improved tricycle rickshaw tires and road asphalt.

C_{rc}, is the rolling resistance coefficient, and its value is assumed as 0.008 for rubber tires and asphalt.

The aerodynamic drag of the improved tricycle is given by Equation (3.4) [5].

$$F_d = \frac{1}{2} C \rho A V^2 \tag{3.4}$$

Where F_d is the aerodynamic drag force (N)

C is drag coefficient (0.8)
ρ, is the density of air (kg/m^3)
A is the frontal area (m^2) of the improved tricycle.
V is the velocity of improved tricycle (m/s)

The grading resistance of an improved tricycle is given by Equation (3.5) [6].

$$F_{gr} = mg Sin\theta \tag{3.5}$$

Where F_{gr} is grading resistance in (Newton)

θ is the angle of road inclination

3.4.3 CHAIN DRIVE ANALYSIS

Driving sprocket is bigger than the driven sprocket as shown in Figure 3.1.
Input torque on sprocket 1 is given by Equation (3.6) [22].

$$T_{in} = x \times l \tag{3.6}$$

Hence, force induced at the chain from the sprocket 1 is given by Equation (3.7)

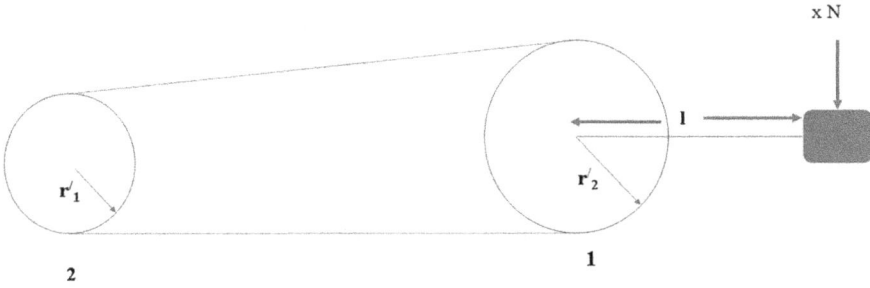

FIGURE 3.1 Driving sprocket is bigger than driven.

$$F = \frac{x \times l}{r_1} \tag{3.7}$$

Output torque at sprocket 2 is given by Equation (3.8) [22].

$$T_{out} = F \times r_2 = \frac{x \times l \times r_2}{r_1} \tag{3.8}$$

Let the total weight of the tricycle, including passengers sitting, is W.
   Therefore, output torque to weight ratio is given by Equation (3.9) [22].

$$\varphi = \frac{T_{out}}{W} = \frac{x \times l \times r_2}{W r_1} \tag{3.9}$$

Now, since, $r_1 > r_2$. Hence, $\frac{r_2}{r_1} < 1$
   In case the driving sprocket is smaller than the driven sprocket as shown in Figure 3.2 below.
   Input torque in sprocket 1 is given by Equation (3.10)

$$T'_{in} = x \times l \tag{3.10}$$

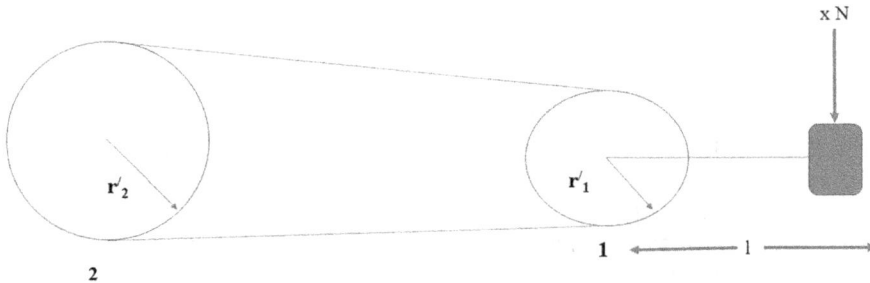

FIGURE 3.2 Driving sprocket is smaller than driven sprocket.

Hence, force generated at the chain from sprocket 1 is given by Equation (3.11) [16].

$$F' = \frac{x \times l}{r'_1} \tag{3.11}$$

Output torque at sprocket 2 is estimated by Equation (3.12)

$$T'_{out} = F' \times r'_2 = \frac{x \times l \times r'_2}{r'_1} \tag{3.12}$$

Therefore, output torque to weight ratio is given by Equation (3.13) [22].

$$\varphi' = \frac{T'_{out}}{W} = \frac{x \times l \times r'_2}{Wr'_1} \tag{3.13}$$

Since, $r'_2 > r'_1$. Hence, $\frac{r'_2}{r'_1} > 1$

$$\therefore \frac{r'_2}{r'_1} > \frac{r_2}{r_1} \tag{3.14}$$

Hence, $\varphi' > \varphi$ and this is torque multiplication.

3.4.4 MATERIAL SELECTION FOR THE DRIVE TRAIN

The transmission mechanism consists of five sprockets of 200 mm diameter, three sprockets of 86 mm, a solid shaft of 23 mm diameter, four standard tricycle rickshaw chains, and the base frame made up of square hollow mild steel bars of 25 mm × 25 mm and thickness of 3 mm. Three sprockets of 200 mm diameter and one sprocket of 86 mm diameter were fixed with the driving shaft. One sprocket of 200 mm diameter was fixed with the pedal. One sprocket of 86 mm diameter and one sprocket of 200 mm were fixed with the right rear wheel hub. And finally, one sprocket of 86 mm diameter was fixed with the left rear wheel hub.

3.4.5 MODELING OF THE HUMAN-POWERED DRIVE TRAIN

Hybrid tricycle's manual, human-driven power train has been mathematically modelled, and the link between applied forces/torques and the vehicle's resulting motion has been studied. Pedaling involves the use of two rigid bodies (links): thigh-link 1 and shin-link 2, which are linked together using kinematic pairings or joints. Figure 3.2 depicts the traditional depictions of the two types of joints, which are believed to be mostly revolute. A kinematic chain connects all of the parts of the construction. The hip (origin) is restricted to a point and the pedal crank is coupled to the end-effector (shin endpoint), which is the kinematic pair's final result. The locations and orientations of the two links are computed in the kinematic model,

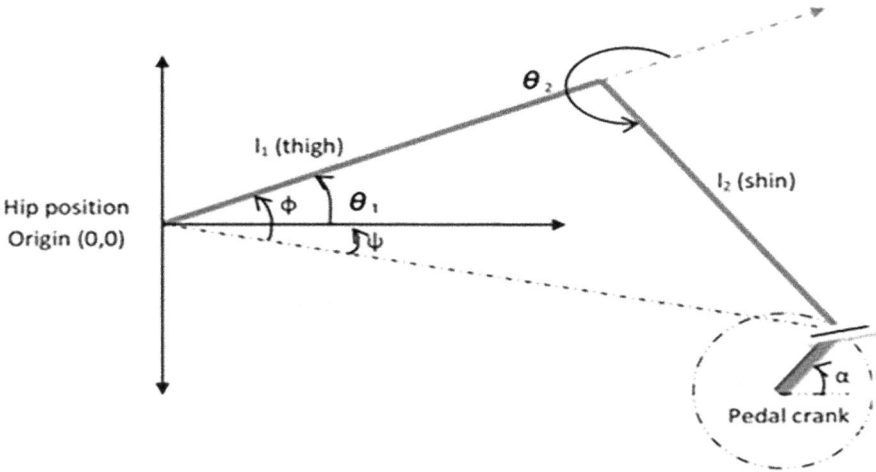

FIGURE 3.3   Joint Representation.

which in this case is an inverse model, utilizing the known positions and orientations of the end-effector, i.e., the pedal crank. The modeling human-powered drive train is described below.

3.4.5.1   Kinematic and Dynamic Modelling of Human Drive System

Figure 3.3 represents the joints of both links, that is Thigh (link1) and Shin (link 2)
Calculating the D-H parameters for the Thigh (link 1) and Shin (link 2) [23]

Link No.	b_i	α_i	l_i	θ_1
1	0	0	l_1	θ_1
2	0	0	l_2	θ_2

The transformation matrix:

$$T_1 = [cos\theta_1 \ \theta_1 \ 0 \ l_1 cos \ \theta_1 \ sin \ sin \ \theta_1 \ cos \ cos \ \theta_1 \ 0 \ l_1 sin \ \theta_1 \ 0 \ 0 \ 1 \ 0 \ 0 \ 0 \ 0 \ 1]$$

Similarly,

$$T_2 = [cos\theta_2 \ - \ sin\theta_2 \ 0 \ l_2 cos\theta_2 \ sin \ sin \ \theta_2 \ cos\theta_2 \ 00000 \ l2sin\theta_2 \ 1001]$$

So the final transformation matrix is: $T = T_1 \times T_2$

$$T = [cos\theta_1 \ \theta_1 \ 0 \ l_1 cos\theta_1 \ sin \ sin \ \theta_1 \ cos \ cos \ \theta_1 \ 0 \ l_1 sin\theta_1 \ 00100001]$$
$$\times [cos\theta_2 \ - \ sin\theta_2 \ 0 \ l_2 cos\theta_2 \ sin \ sin \ \theta_2 \ cos\theta_2 \ 00000 \ l2sin \ \theta_2 \ 1001]$$

For an inverse problem

$$T^I = [cos\Phi - sin\Phi \ 0 \ P_x \ sin\Phi \ cos\Phi \ 00000 \ P_y 1001]$$

Comparing two equations,

$$\Phi = \theta_1 + \theta_2$$

$$P_x = l_1 \cos \theta_1 + l_2 \cos(\theta_1 + \theta_2) \tag{3.15}$$

$$P_y = l_1 \sin \theta_1 + l_2 sin (\theta_1 + \theta_2) \tag{3.16}$$

Now the position:

$$P_x = b + R \ cos\alpha \ \& \ P_y = a - R \ sin\alpha \tag{3.17}$$

Using (15) in (16) and (17)

$$(b + R \ cos\alpha) = l_1 cos\theta_1 + l_2 cos\Phi$$

$$(a - R \ sin\alpha) = l_1 sin\theta_1 + l_2 sin\Phi$$

$$tan\Phi = \frac{P_y}{P_x} = \frac{(a - R \ sin\alpha)}{(b + R \ cos\alpha)}$$

So,

$$\Phi = tan^{-1}\left(\frac{(a - R \ sin\alpha)}{(b + R \ cos\alpha)}\right)$$

From Geometry,

$$\theta_2 = -cos^{-1}(R^2 - l_1^2 - l_2^2/2 \ l_1 l_2)$$

$$\theta_1 = 360 + [\Phi - \Psi]$$

Where, $\Psi = 360 - cos^{-1}\left(\frac{R^2 + l_1^2 - l_2^2}{2 \ Rl_1}\right)$
And

$$R = \sqrt{P_x^2 + P_y^2} \tag{3.18}$$

So,

$$\theta_1 = \tan^{-1}\left(\frac{(a - R\sin\alpha)}{(b + R\cos\alpha)}\right) + \cos^{-1}\left(\frac{b^2 + R^2 + a^2 + 2bR\cos\alpha - 2aR\sin\alpha + l_1^2 - l_2^2}{2l_1\sqrt{((b + R\cos\alpha)^2 + (a - R\sin\alpha)^2)}}\right).$$

(3.19)

And

$$\theta_2 = \cos^{-1}\left(\frac{b^2 + R^2 + a^2 + 2bR\cos\alpha - 2aR\sin\alpha + l_1^2 - l_2^2}{2l_1 l_2}\right)$$

3.4.5.2 Simulating the Kinematic Model

MATLAB was used to construct and simulate the inverse kinematic issue for the investigated kinematic pair. The kinematic model's output was used to help develop the dynamic model.

The findings of the kinematic modelling showed the different locations and orientations of the thigh, link-1, and the shin, link-2. Figure 3.4 depicts the thigh and shin positions in relation to the pedal crank action. Figures 3.5 and 3.6 show the relationship between the thigh and shin angles as a function of pedal crank angle in degrees. According to research, the thigh and shin angle changes in sinusoidal pattern with the pedal crank angle.

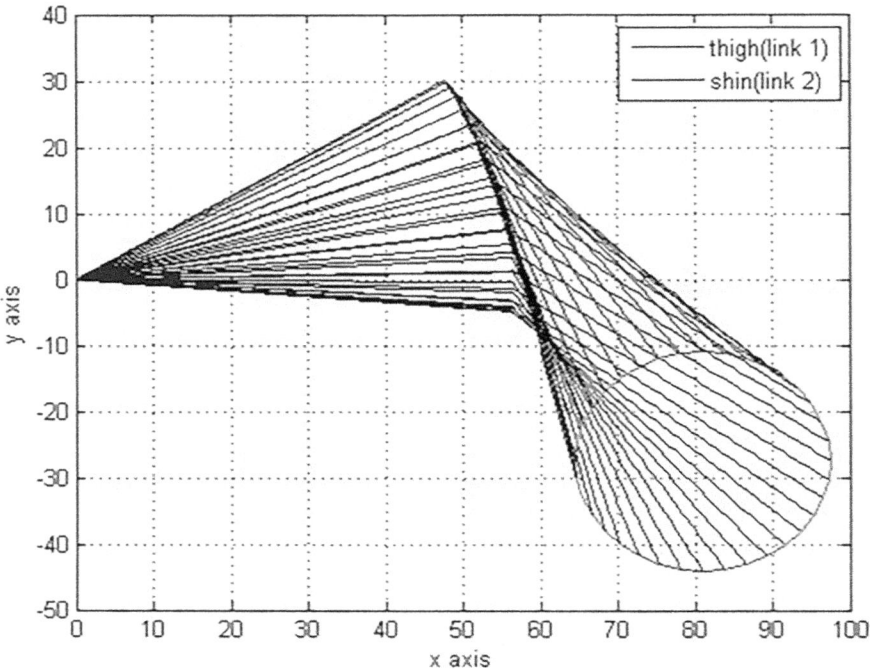

FIGURE 3.4 Position and orientation of the links.

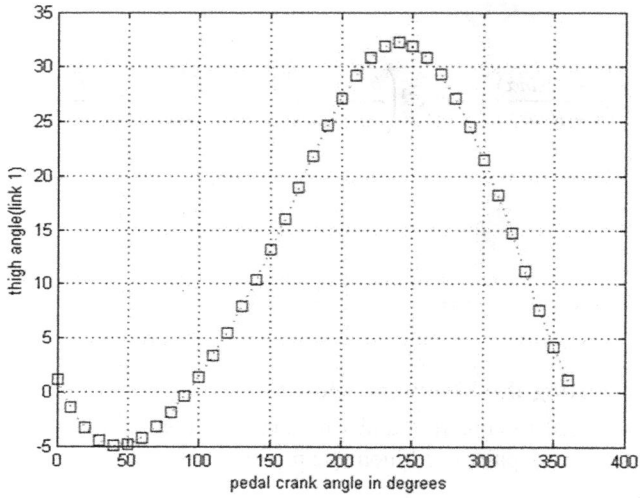

FIGURE 3.5 Variation of thigh angle vs. pedal crank.

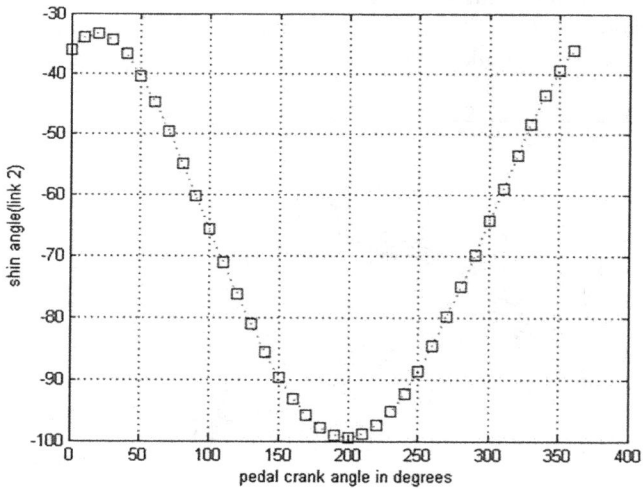

FIGURE 3.6 Variation of shin angle vs. pedal crank.

3.4.5.3 Dynamic Modelling

Similarly, to kinematics, there are two potential "models" for dynamics:

- **Direct model:** Once the forces/torques applied to the joints, as well as the joint positions and velocities, are known, we can compute the joint accelerations.

$$\tau(t) = >\ddot{\theta}(t),\ \dot{\theta}(t),\ \theta(t)$$

- **Inverse model:** Once the joint accelerations, velocities, and positions are known, we can compute the corresponding forces/torques.

$$\ddot{\theta}(t),\ \dot{\theta}(t),\ \theta(t) = >\tau(t)$$

There are two approaches to the definition of the dynamic model, they are:

- **EULER-LAGRANGE approach:** Because it is simpler and more logical, this dynamic model is more suited for understanding the consequences of changes in mechanical parameters. The linkages are considered collectively, and the model has derived analytically. For the model, the kinetic and potential energy are used (which is not intuitive), and the derivatives with respect to the parameters are used to account for its degree of freedom.
- **NEWTON-EULER approach:** It utilizes the serial nature of an industrial manipulator to implement a computationally efficient recursive method. The mathematical model, on the other hand, is not expressed in enclosed form. It's a tie between the two methods.

We know that:

[1] The Kinetic Energy function is K $(\dot{\theta}(t), \theta(t))$.
[2] The Potential Energy function is P $(\theta(t))$.

And therefore, The Lagrangian function

$$L(\dot{\theta}, \theta) = K(\dot{\theta}, \theta) - P(\theta)$$

The Euler-Lagrange equations are:

$$\Psi i = \frac{d}{dt}\left(\frac{\partial L}{\partial \dot{\theta} i}\right) - \frac{\partial L}{\partial \theta i}, \quad i = 1, 2 ..., \text{n} \qquad (3.20)$$

Ψi being the non-conservative (external or dissipative) generalized forces performing work on θ_i.

Since the potential energy does not depend on the velocity, the Euler-Lagrange equations can be rewritten as:

$$\Psi i = \frac{d}{dt}\left(\frac{\partial K}{\partial \dot{\theta} i}\right) - \frac{\partial K}{\partial \theta i} + \frac{\partial P}{\partial \theta i}, \quad i = 1, 2. ..., \text{n} \qquad (3.21)$$

It is dynamic modelling that considers inertia forces, Coriolis components, gravitational forces, and forces acting on the end effector. The tractive force of the vehicle is taken into account while calculating the end effector force. The moment at the axle determines the overall tractive force that changes with the drivetrain. Knowing the axle's tractive effort gives us the pedal crank moment.

End effector forces include tangential and radial components. The Euler-Lagrangian approach is used to study dynamic systems dynamically (classical approach using Inverse model). The total kinetic energy of the thighs and shins (links 1 and 2) is computed. Like the potential energy, it's likewise computed based on location alone and is not affected by speed.

The Lagrange's equation is:

$$L = (K1 + K2) - (P1 + P2) \tag{3.22}$$

And

$$\Psi i = \frac{d}{dt}\left(\frac{\partial L}{\partial \dot{\theta} i}\right) - \frac{\partial L}{\partial \theta i}, \quad \text{i = 1, 2 ..., n is calculated.}$$

There are also torques owing to gravity and Coriolis, as well as moments due to an end effector. This produces the actuation torques, which are equal to these four forces.

3.4.6 Dynamic Model of Thigh (link 1) and Shin (link 2)

Considering two degrees of freedom with joint variables θ_1 and θ_2.

The kinetic energy (K) and potential energy (P) for thigh (link 1),

$$K_1 = \tfrac{1}{2}\ m_1 l_{c1}^2 \dot{\theta}_1^2 + \tfrac{1}{2} I_1 \dot{\theta} l_1^2.$$

$$P_1 = m_1 g l_{c1}^2 \theta_1$$

Position of the end-effectors,

$$P_x = l_1 cos\ cos\ \theta_1 + l_2 cos\ cos\ (\theta_1 + \theta_2)$$

$$P_y = l_1 sin\ sin\ \theta_1 + l_2 sin\ sin\ (\theta_1 + \theta_2)$$

Also, the position of the Centre of Mass,

$$P_{cx} = l_1 cos\ cos\ \theta_1 + l_{c2} cos\ cos\ (\theta_1 + \theta_2)$$

$$P_{cy} = l_1 sin\ sin\ \theta_1 + l_{c2} sin\ sin\ (\theta_1 + \theta_2)$$

Velocity of the Centre of Mass,

$$\dot{P}_{cx} = (-l_1 sin\ sin\ \theta_1)\ \dot{\theta}_1 - l_{c2} sin\ sin\ (\theta_1 + \theta_2).\ (\dot{\theta}_1 + \dot{\theta}_2)$$

$$\dot{P}_{cy} = (l_1 cos\ cos\ \theta_1)\dot{\theta}_1 + l_{c2}cos\ cos\ (\theta_1 + \theta_2).\ (\dot{\theta}_1 + \dot{\theta}_2)$$

Now, Kinetic Energy (K) and potential Energy (P) of the shin (link 2),

$$K_2 = \tfrac{1}{2}\ m_2\dot{P}_c^T\ \dot{P}_c + \tfrac{1}{2}\ I_2(\dot{\theta}_1 + \dot{\theta}_2)^2$$

$$P_2 = m_2g\ [l_1 sin\ sin\ \theta_1\dot{\theta}_1 + l_{c2}sin\ sin\ (\theta_1 + \theta_2)]$$

Now, $\dot{P}_c^T\ \dot{P}_c = l_1^2\dot{\theta}_1^2 + l_{c2}^2(\dot{\theta}_1 + \dot{\theta}_2)^2 + 2\ l_1 l_{c2}cos\ cos\ \theta_2(\dot{\theta}_1^2 + \dot{\theta}_1\dot{\theta}_2)$
Now,

$$L = (K_1 + K_2) - (P_1 + P_2)$$
$$= \left[\tfrac{1}{2}\ m_1 l_{c1}^2\dot{\theta}_1^2 + \tfrac{1}{2}\ I_1\dot{\theta}_1^2\right] + \left[\tfrac{1}{2}\ m_2 l_1^2\dot{\theta}_1^2 + 1/2m_2 l_{c2}^2(\dot{\theta}_1 + \dot{\theta}_2)^2 \right.$$
$$\left. + l_1 l_{c2}cos\ cos\ \theta_2(\dot{\theta}_1^2 + \dot{\theta}_1\dot{\theta}_2)\right]$$
$$+ \tfrac{1}{2}\ I_2(\dot{\theta}_1 + \dot{\theta}_2)^2 - m_1 gl_{c1}^2 sin\ sin\ \dot{\theta}_1$$
$$- m_2g\ [l_1 sin\ sin\ \theta_1\dot{\theta}_1 + l_{c2}sin\ sin\ (\theta_1 + \theta_2)]$$

$$\frac{\partial L}{\partial \theta_1} = -m_1 gl_{c1}cos\ cos\ \theta_1 - m_2gl_1 cos\ cos\ \theta_1 - m_2gl_{c2}cos\ cos\ (\theta_1 + \theta_2)$$

$$\frac{\partial L}{\partial \theta_2} = -m_1 l_1 l_{c2}sin\ sin\ \theta_2\left(\dot{\theta}_1^2 + \dot{\theta}_1\dot{\theta}_2\right) - m_2g\ l_{c2}cos\ cos\ (\theta_1 + \theta_2)$$

$$\frac{\partial L}{\partial \dot{\theta}_1} = m_1 l_{c1}^2\dot{\theta}_1 + I\dot{\theta}_1 + m_2 l_1^2\dot{\theta}_1 + m_2 l_{c2}^2(\dot{\theta}_1 + \dot{\theta}_2) + 2m_2 l_1 l_{c2}\ cos\ cos\ \theta_2\dot{\theta}_1$$
$$+ m_2 l_1 l_{c2}\ cos\ cos\ \theta_2\dot{\theta}_2 + I_2(\dot{\theta}_1 + \dot{\theta}_2)$$

$$\frac{\partial L}{\partial \dot{\theta}_2} = m_2 l_{c2}^2(\dot{\theta}_1 + \dot{\theta}_2) + m_2 l_1 l_{c2}\ cos\ cos\ \theta_2\dot{\theta}_1 + m_2 l_1 l_{c2}\ cos\ cos\ \theta_2\dot{\theta}_2 + I_2(\dot{\theta}_1 + \dot{\theta}_2)$$

$$\frac{d}{dt}\left(\frac{\partial L}{\partial \dot{\theta}_1}\right) = m_1 l_{c1}^2\ddot{\theta}_1 + I_1\ddot{\theta}_1 + m_2 l_1^2\ddot{\theta}_1 + m_2 l_{c2}^2(\ddot{\theta}_1 + \ddot{\theta}_2) + 2\ m_2 l_1 l_{c2}\ cos\ cos\ \theta_2\ddot{\theta}_1$$
$$- 2\ m_2 l_1 l_{c2}\ sin\ sin\ \theta_2\ddot{\theta}_1\dot{\theta}_2 + m_2 l_1 l_{c2}\ cos\ cos\ \theta_2\ddot{\theta}_2 + I_2(\ddot{\theta}_1 + \ddot{\theta}_2)$$

$$\frac{d}{dt}\left(\frac{\partial L}{\partial \dot{\theta}_2}\right) = m_2 l_{c2}^2 (\ddot{\theta}_1 + \ddot{\theta}_2) + m_2 l_1 l_{c2} \cos \cos \theta_2 \ddot{\theta}_1 - m_2 l_1 l_{c2} \sin \sin \theta_2 \dot{\theta}_1 \dot{\theta}_2$$

Now,

$$t_1 = \frac{d}{dx}\left(\frac{\partial l}{\partial \dot{\theta}_1}\right) - \left(\frac{\partial l}{\partial \theta_1}\right) \text{ and } t_2 = \frac{d}{dt}\left(\frac{\partial l}{\partial \dot{\theta}_2}\right) - \left(\frac{\partial l}{\partial \theta_2}\right)$$

$$
\begin{aligned}
\tau_1 &= (m_1 l_{c1}^2 + I_1 + m_2 l_1^2 + m_2 l_{c2}^2 + 2 m_2 l_1 l_{c2} \cos \theta_2 + I_2) \ddot{\theta}_1 \\
&\quad + (m_2 l_{c2}^2 + m_2 l_1 l_{c2} \cos \theta_2 + I_2) \ddot{\theta}_2 + (m_1 g l_{c1} \cos \theta_1 + m_2 g l_1 \cos \theta_1 \\
&\quad + m_2 g l_{c1} \cos(\theta_1 + \theta_2)) \\
&\quad - (2 m_2 l_1 l_{c2} \sin \theta_2 \ddot{\theta}_1 \ddot{\theta}_2) - m_2 l_1 l_{c2} \sin \theta_2 \dot{\theta}_2^2 + rxF_1
\end{aligned}
\tag{3.23}
$$

$$
\begin{aligned}
\tau_2 &= (m_2 l_{c2}^2 + m_2 l_1 l_{c2} \cos \theta_2 + I_2) \ddot{\theta}_1 + (m_2 l_{c2}^2 + I_2) \ddot{\theta}_2 + m_2 l_1 l_{c2} \sin \theta_2 \dot{\theta}_1^2 \\
&\quad + m_2 g l_{c2} \cos(\theta_1 + \theta_2) + rxF_2
\end{aligned}
\tag{3.24}
$$

The above two equations, 23 and 24, can be represented in matrix form as:

$$\tau = [\tau_1 \tau_2] \tag{3.25}$$

3.4.6.1   Inertia Torque Matrix

$$
\begin{aligned}
M &= [m_1 l_{c1}^2 + I_1 + m_2 l_1^2 + m_2 l_{c2}^2 + 2 m_2 l_1 l_{c2} \cos\theta_2 + I_2 m_2 l_{c2}^2 + m_2 l_1 l_{c2} \cos\theta_2 \\
&\quad + I_2 m_2 l_{c2}^2 + m_2 l_1 l_{c2} \cos\theta_2 + I_2 m_2 l_{c2}^2 + I_2]
\end{aligned}
$$

3.4.6.2   Gravitational Torque Matrix

$$G = [m_1 g l_{C1} \cos\theta_1 + m_1 g l_1 \cos\theta_1 + m_2 g l_{C1} \cos(\theta_1 + \theta_2) m_2 g l_{C2} \cos(\theta_1 + \theta_2)]$$

3.4.6.3   Coriolis Torque Matrix

$$V = \left[-(2 m_2 l_1 l_{c2} \sin\theta_2 \ddot{\theta}_1 \ddot{\theta}_2) - m_2 l_1 l_{c2} \sin\theta_2 \dot{\theta}_2^2 \, m_2 l_1 l_{c2} \sin\theta_2 \dot{\theta}_1^2 \right]$$

3.4.6.4   End Effector Force Torque matrix

$$F = [r \times F_1 r \times F_2] \tag{3.26}$$

3.4.7 SIMULATION OF THE DYNAMIC MODEL

Assumptions-

- The thigh and the shin are considered as links.
- The shin and the foot are considered as a single link.
- The Centre of Gravity of both links is considered at their midpoints.
- The angular velocity of the pedal crank is considered to be constant.
- Pedal force assumptions-
- The position of the pedal is assumed to be always at an angle of 30° in the positive direction of the X-axis

3.4.6.5 Modelling Parameters

Following are the parameters which are considered

- Mass of thigh (link 1)= 3.54375 kg
- Mass of shin (link 2) = 2.08575 kg
- Length of thigh (link 1) = 0.565 m
- Length of shin (link 2) = 0.50 m
- Length of pedal crank = 165 mm

Figure 3.7 shows the free-body characteristics of the forces operating on the pedal and crank1.

N is the typical pedal force that acts. The final effector strength is F, composed of two parts: F_r and F_t, both radial and tangential. It is determined through our experimental work and is also expected that the pedal creates an angle of 300° with the positive direction of the X-axis during the circular motion of 360° of the crank. Thus, during the pedaling cycle, the direction of the normal force stays constant.

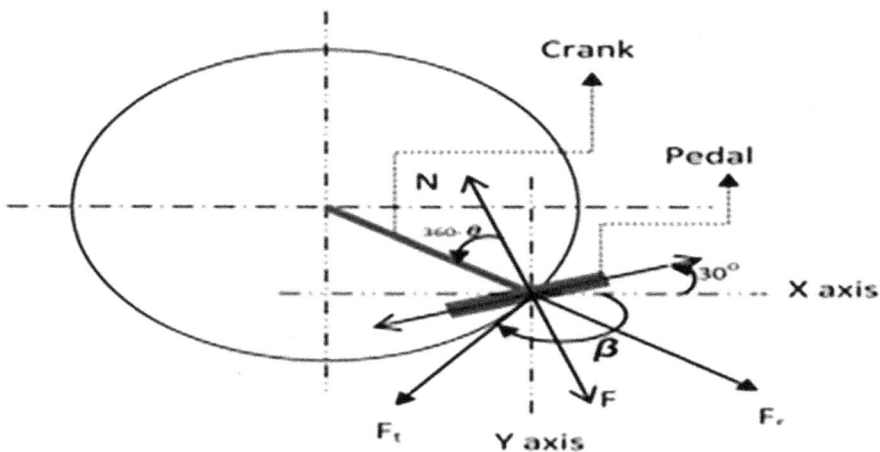

FIGURE 3.7 Pedal force direction.

As the tangential force alone is responsible for the moment at the end effector and its value being known from the total tractive force, the radial component of the normal force is hence neglected. The Figure 3.7 represents the pedal force direction

The instantaneous pedal torque is hence found out by the formula [23–26]

$$\Gamma_{pedal} = (F_x \cos \theta_2 - F_y \sin \theta_2)L_c \qquad (3.27)$$

Here, F_x and F_y are the horizontal and the vertical components of the tangential component of the pedal force whose magnitude is known.

Here $\theta_{.2}$ is the angle between the crank arm and the normal to the pedal surface.

Also, as per the configuration of the vehicle in the study, the initial position of the leg at the beginning of a power stroke is shown in Figure 3.7

For simulation, three cases were considered taking three different angular velocities of the pedal crank.

3.4.6.6 Power Stroke Considerations

One power stroke is considered as 180° of pedal crank rotation. As per the configuration of the vehicle in the study, and a few experimental observations, the power stroke is considered to start from a crank angle of 45° to 225° and again for the second leg from a crank angle of 225° to 405°.

3.5 RESULTS AND DISCUSSION

3.5.1 SIMULATIONS

3.5.1.1 Case I

For the 1st case, one power stroke for the first legs is considered at three different vehicle speeds.

$$V_1 = 6.3 \; km/h, \quad V_2 = 11.6 \; km/h, \quad V_3 = 17 \; km/h$$

Accordingly, the torque at the end effector, considering the tangential force at the pedal crank, comes out to be-

$$\Gamma_1 = 13.35 \; Nm, \quad \Gamma_2 = 8.00 \; Nm, \quad \Gamma_3 = 2.66, \; Nm \; respectively.$$

The inverse kinematic problem for the kinematic pair in the study was mathematically modelled and simulated in MATLAB. The results obtained from the kinematic model have been later utilized to formulate the dynamic model.

The above plots (from Figure 3.8 to 3.12) show the variation of different parameters vs. the crank angle and time.

One power stroke for the first leg is considered at three different vehicle speeds. While considering the leg, Thigh (link-1) and Shin (link-2) are considered separately. It is seen that the torque required at higher speed is less as compared to lower speeds for a particular crank angle.

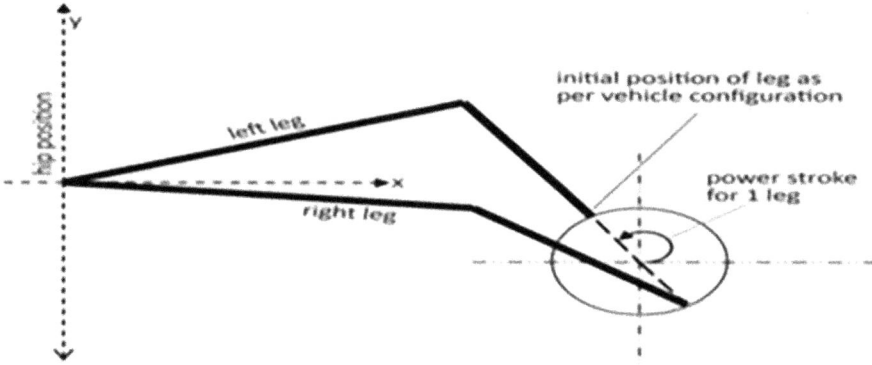

FIGURE 3.8 Initial position of pedal cranks.

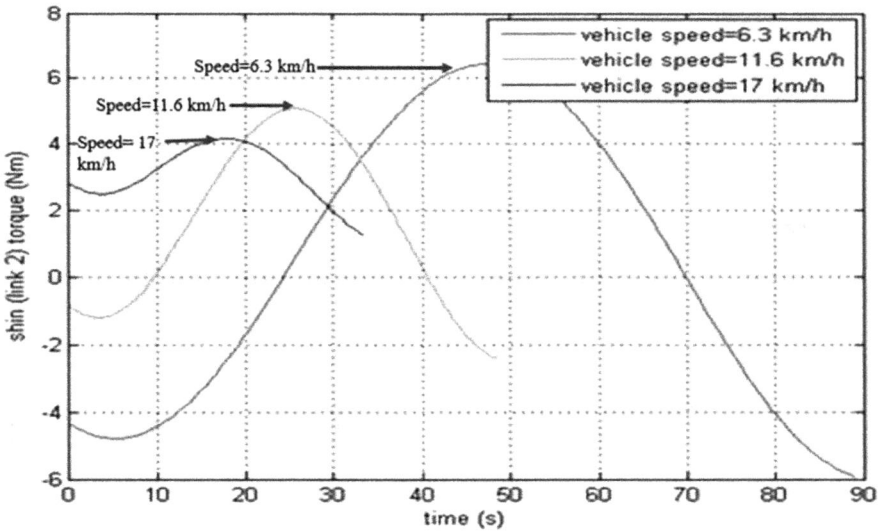

FIGURE 3.9 Thigh Torque.

3.5.1.2 Case II

For the 2nd case, the power strokes for both the legs are considered one after the other keeping the speed at the wheels fixed at 30 rpm.

Accordingly, the torque at the end effector, considering the tangential force at the pedal crank, comes out to be-

$$\Gamma_1 = 13.2 \ N\text{-}m$$

Figure 3.13 shows the velocity of shin. Figure 3.14 to 3.19 illustrate the time history and crank angle locations of link with the velocity, acceleration, and torque

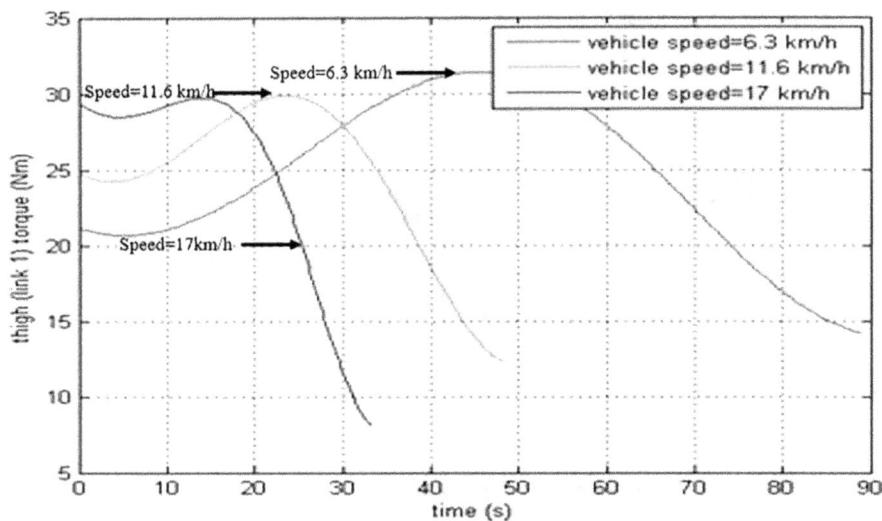

FIGURE 3.10   Shin Torque.

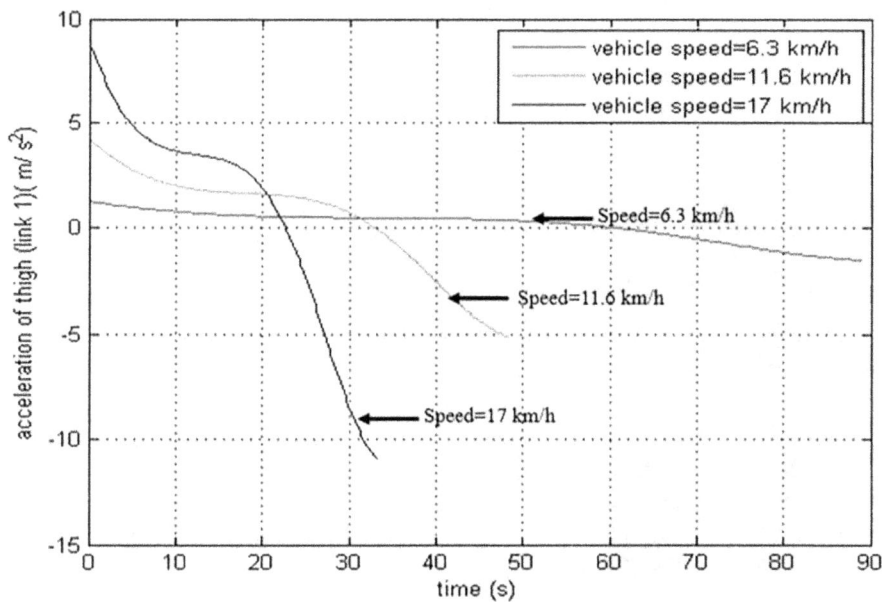

FIGURE 3.11   Acceleration of Thigh.

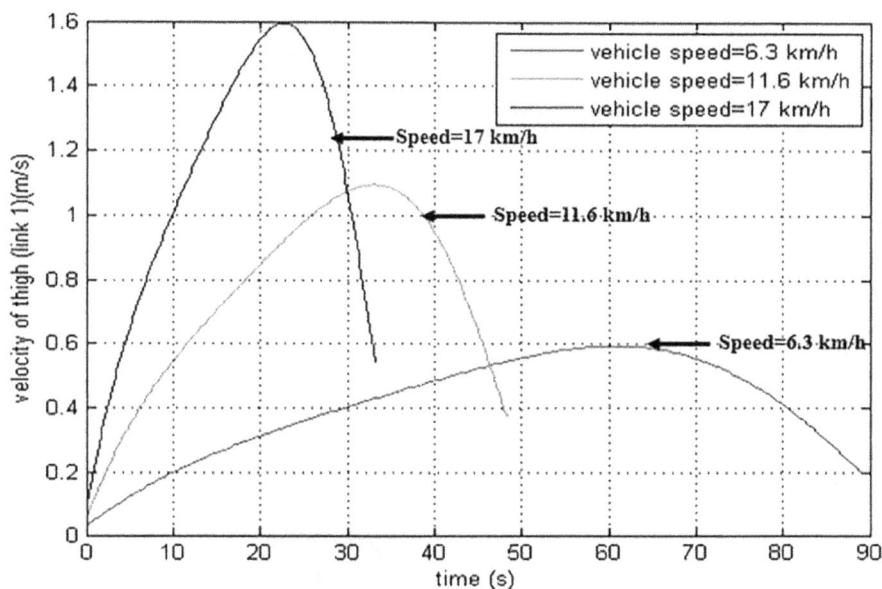

FIGURE 3.12   Velocity of Thigh.

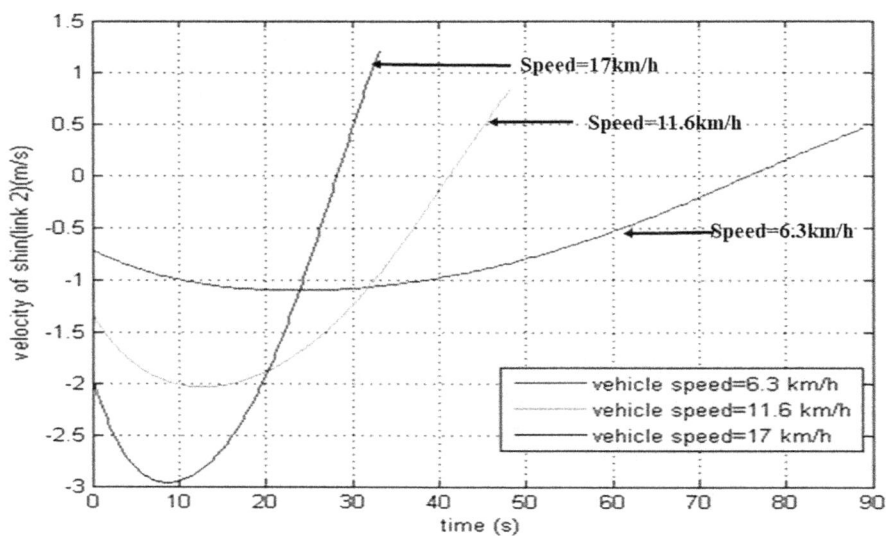

FIGURE 3.13   Velocity of Shin.

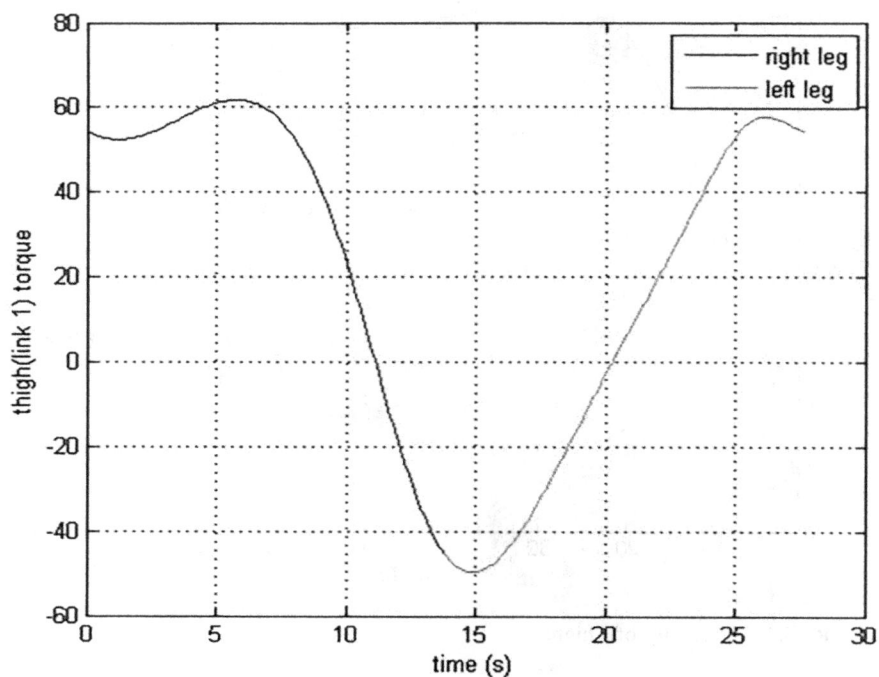

FIGURE 3.14   Thigh Torque.

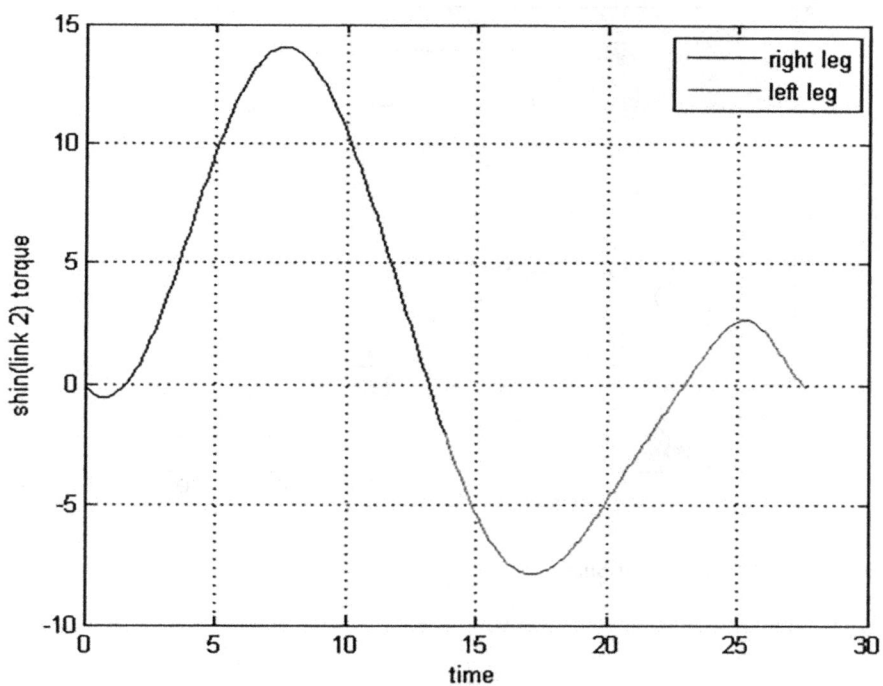

FIGURE 3.15   Shin Torque.

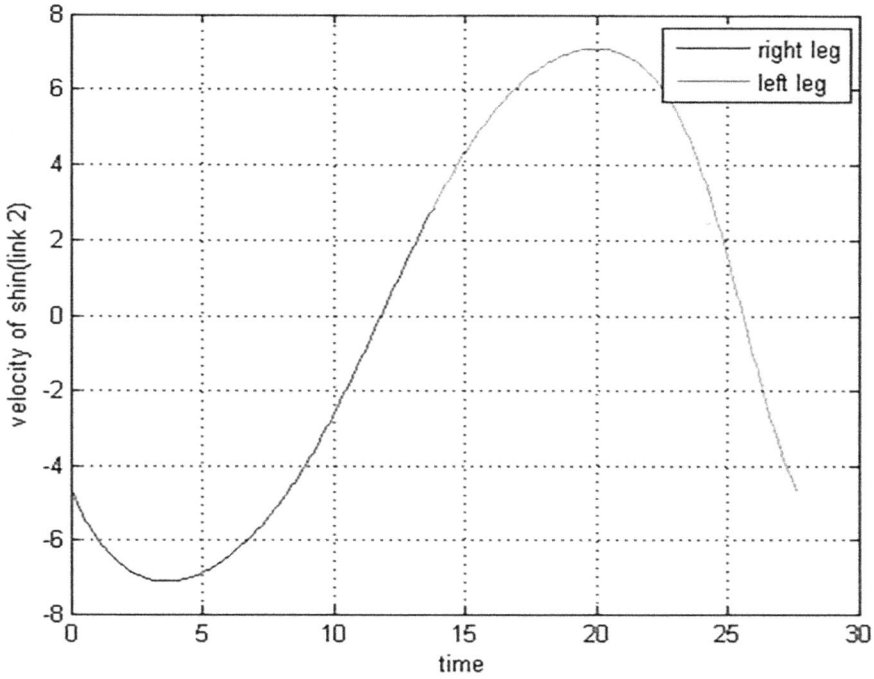

FIGURE 3.16   Shin Velocity.

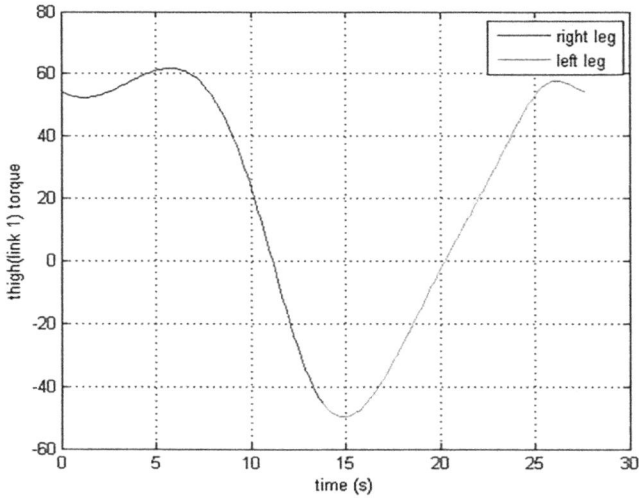

FIGURE 3.17   Velocity of Shin.

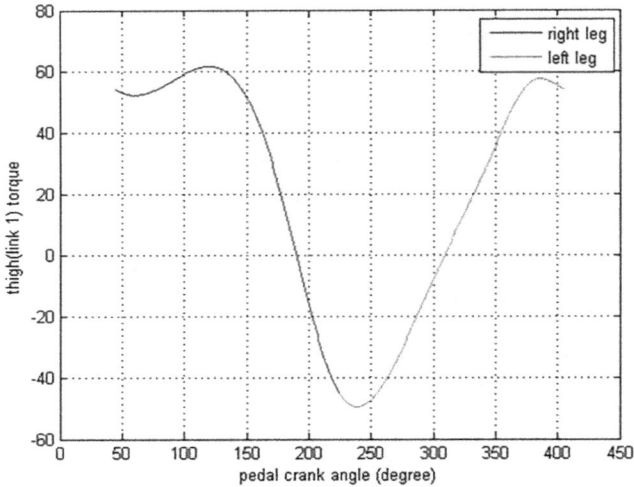

FIGURE 3.18　Velocity of Thigh.

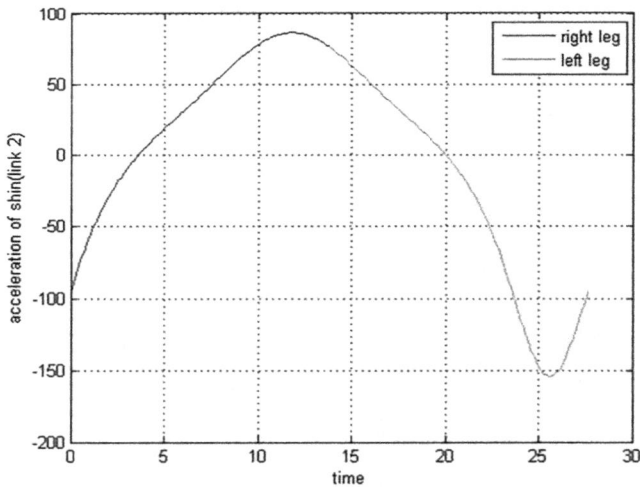

FIGURE 3.19　Acceleration of Shin.

resulting from the joint trajectories for both the legs with symmetrical velocity profiles and equal time duration. Each joint makes a revolution of 180 degrees in a span of about 14 seconds. The velocity and acceleration of the thigh (link-1) are compared with time. The velocity increases up to 9 seconds for the first power stroke and the corresponding slope gives the acceleration that decreases from the start and reaches a zero value at nearly 9 seconds from the

The velocity and acceleration are plotted against the time for shin (link-2). The velocity first decreases and then increases to reach a maximum at nearly 14 seconds for the first power stroke. The corresponding slope gives the acceleration that increases from the start and reaches a zero value at nearly 4 seconds from the start.

The torque of the inertia of joint 1 follows the acceleration history. The inertia torque is somewhat consistent in joint 2 owing to joint 2 acceleration since the moment of inertia is consistent in joint 2 axis is constant. The inertia torque at each joint due to acceleration of the other joint confirms the symmetry of the inertia matrix, since the acceleration profiles are the same for both joints. The thigh provides the maximal inertia torque at the beginning, but the brightness torque continues to grow until it reaches a maximum value of around 14 seconds.

The Coriolis effect is only visible on joint 1 since the shine moves in relation to the thigh movable frame (link 1) but is stationary on the shin frame (link-2). At around 2.5 seconds from the first power stroke, the thigh produces a maximum torque of Coriolis, and the glow is 9 seconds.

The total torque is plotted against the time for both the links. The thigh gives the maximum torque at nearly 6 seconds from the start of the stroke, while that for the shin is around 7.5 seconds.

3.5.2 Modelling of Tractive Effort

Table 3.1 is the maximum traction force required for five passengers with an average 60 kg mass is 485.98 N at an average velocity of 1m/s and a minimum of 105.57 N without any passenger at velocity of 2.25 m/s. If we neglect grading resistance, then the corresponding load is much lesser, i.e., 31.07 N and 9.8 N. This load is much lesser for rickshaw pullers. However, for encountering slope (0.12 radian), the bigger gear (sprocket) with a gear ratio of 2.5 may be used effectively.

3.5.3 Working Mechanism of Improved Tricycle

At starting, the rickshaw puller pedals towards the reverse way, sprocket (2) is involved and rotates in the reverse way. This causes the power to flow towards

TABLE 3.1
Traction Loads Modeling

W (kg)	F_r(N)	V (m/s)	F_d (N)	F_{gr}(N)	$F_r + F_{gr}$ (N)	F_T (N)
3800	30.4	1	0.672	454.91	31.07	485.98
3200	25.6	1.25	1.05	383.08	26.65	409.73
2600	20.8	1.5	1.512	311.25	22.31	333.56
2000	16	1.75	2.058	239.42	18.06	257.48
1400	11.2	2	2.688	167.60	13.88	181.48
800	6.4	2.25	3.402	95.77	9.8	105.57

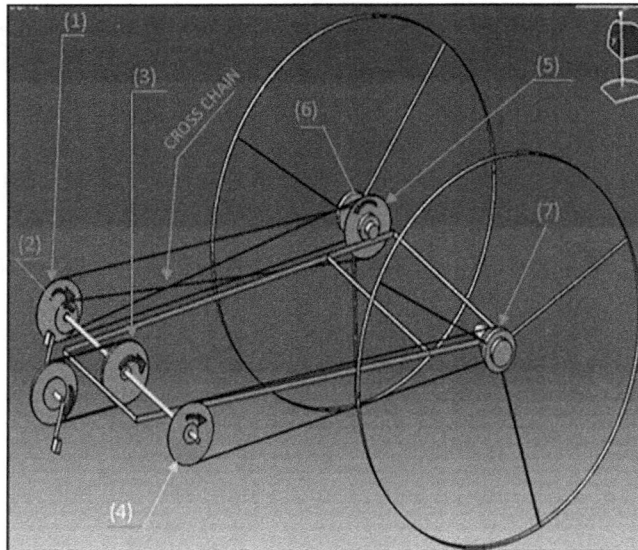

FIGURE 3.20 Pedalling backward direction at starting.

sprocket (5) in an onward way. since a cross chain links sprocket (2) and (5). After attaining enough torque and speed, when the driver starts to pedal in the forward direction, sprockets (1) and (4) are engaged, which in turn transmits the pedaling power to smaller sprockets (6) and (7) respectively by a straight chain drive. When sprocket (2) is engaged at the beginning of pedaling, the torque required for pedaling is very low, and the output torque that we obtain from sprocket (5) is high. Both the operations for starting and uniform the speed of improved tricycle rickshaw are shown in Figure 3.20 and Figure 3.21.

3.5.4 ANALYSIS OF TRICYCLE FRAME

Figure 3.22 represents stress analysis of the base frame of the improved tricycle. It is clear from the figure that the minimum stress is 6.69×10^6 N/m^2 to a maximum of 7.71×10^7 N/m^2 at the rear part of the frame. Similarly, Figure 3.23 represents the stress analysis of the intermediate shaft. The stress variation ranges from 1.75×10^8N/m^2 right-hand part of the shaft to a maximum of 1.89×10^{10} N/m^2 at left-hand part of the shaft. For both cases, the stress ranges are much within the safer limit for a particular loading condition. Figure 3.23 represents the stress analysis of the intermediate shaft.

3.6 FABRICATION OF IMPROVED CYCLE

The Figure 3.24 represents the different stages that have been used to fabricate the improved tricycle.

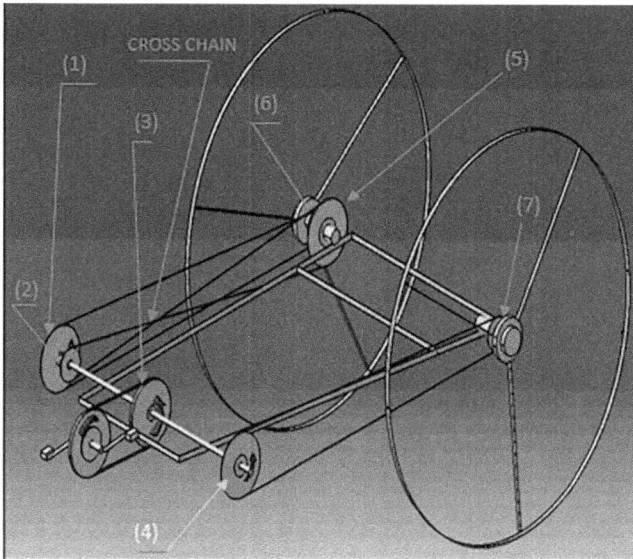

FIGURE 3.21 Pedalling forward direction at uniform speed.

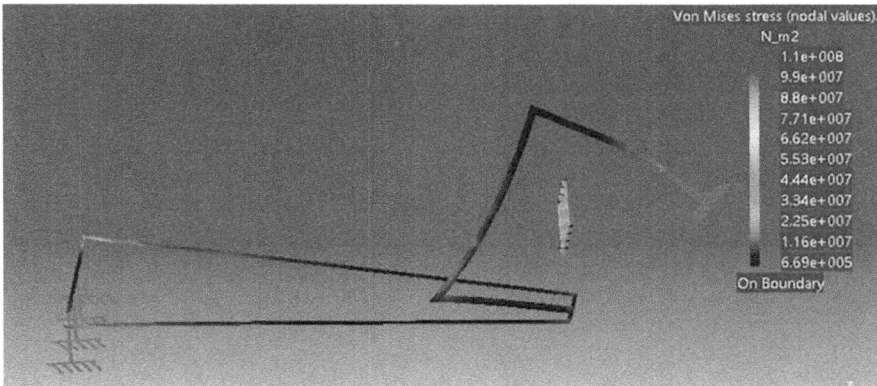

FIGURE 3.22 Stress analysis a base frame.

3.6.1 FABRICATION OF BASE FRAME

Base frame has the following technical specifications. They are

Step 1. Two mild steel square pipes of 1500 mm length, one 600 mm length, and one 900 mm length are cut using a power hacksaw.

Step 2. All of the square pipes cut in the above step are joined by welding like a shape of a trapezium.

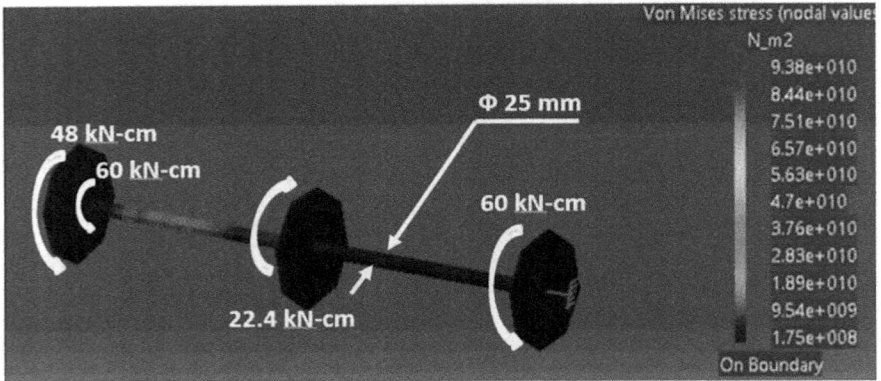

FIGURE 3.23   Stress analysis of the intermediate shaft.

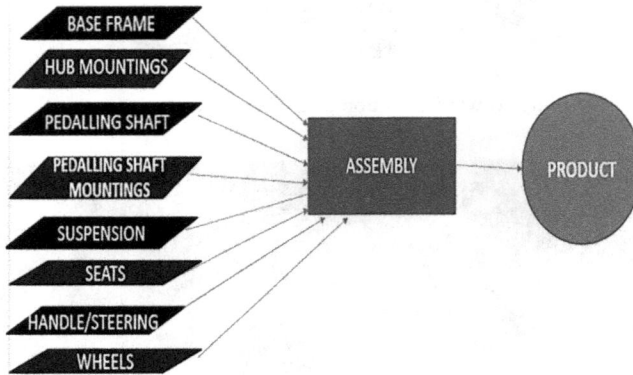

FIGURE 3.24   Fabrication stages of the improved tricycle.

Step 3. A square pipe of 750 mm is cut and welded at a length of 1000 mm from the smaller parallel side of the trapezium. The fabrication of the base frame is represented in the Figure 3.25.

3.6.2   HUB MOUNTINGS

Hub mountings are parts where the hubs of the wheels are attached. For the fabrication of hub mountings, below mentioned steps were followed

Step 1. Two cylindrical solid bars of diameter 52 mm and 68 mm in length are cut from feedstocks.

Step 2. They are machined as shown in Figure 3.26 below.

FIGURE 3.25 Base frame of improved tricycle.

FIGURE 3.26 Manufactured hubs for rear wheels.

3.6.3 HUB

The hub is directly attached to the spokes of the wheel. The hub will rotate along with the wheel.

3.6.4 INTERMEDIATE SHAFT

The Intermediate shaft consists of three sprockets of 20 cm diameter and one Sprocket of 8 cm diameter. Two Sprockets of 20 cm diameter are fixed at extreme ends, one Sprocket of 20 cm diameter at the center of the Intermediate shaft, and one of 8 cm diameter is fixed next to one of the Sprocket at the extreme end.

3.6.4.1 Calculations Involved in Finding the Dimension of Intermediate Shaft

- **Material of the shaft:** Mild Steel
- **Outer Diameter of Shaft** = 25 mm
- **Let Internal Diameter of Shaft** = $2r_1$

$$\therefore 800 \times 28 = F_1 \times 10 \tag{3.28}$$

$$F_1 = 2240 \; N$$

$$\begin{aligned} \therefore T_1 &= F_1 \times 10 = 22400 \; N\text{-}cm \\ T_4 &= T_6 = 60000 \; N\text{-}cm \\ T_3 &= 48000 \; N\text{-}cm \end{aligned} \tag{3.29}$$

Now, By applying Torsion Formula

$$\frac{T}{J} = \frac{G\theta}{L} \tag{3.30}$$

Where,

T = Torque (N-m)
J = Polar Moment of Inertia
G = Modulus of Rigidity
θ = Angle of Twist
L = Length of the shaft (m)

We get,

$$\theta = \frac{1.182}{156.25 - r^2} \tag{3.31}$$

If Intermediate Shaft has to experience minimum Torsional Bending, then

$$\frac{d\theta}{dr_1} = 0 \tag{3.32}$$

Then the value of r_1 comes around 10 m (Approx.)

3.6.4.2   Intermediate Shaft Mountings

Two Housing Bearings (FKD P205) are used to mount the Intermediate shaft with the Base Frame.

3.6.5   REAR WHEEL

The two rear wheels are taken from the Cycle van. The rear wheels of the cycle van are taken because it has a larger ability to carry load compared to the wheels used in other bicycles.

3.6.6   CHAIN OF THE TRICYCLE

On the hub of the left rear wheel, one sprocket of 8 cm diameter (Which works on Ratchet Mechanism) and one sprocket of 20 cm diameter is fixed.

One chain connects the smaller sprocket of 8 cm diameter fixed with the hub of the left rear wheel to the larger sprocket of 20 cm diameter fixed with the intermediate shaft. Another chain connects with the bigger sprocket of 20 cm diameter fixed with the hub of the left rear wheel to the smaller sprocket of 8 cm diameter fixed with the intermediate shaft by cross-chain mechanism.

3.6.7   PASSENGER SEATS

There are four Passenger Seats mounted on the Base Frame by Hydraulic Shock Absorbers and square ducts, whose alignments and seating angle can be adjusted as per the convenience of the Passengers.

The Hydraulic shock Absorbers are added to provide comfort to the passengers by reducing the jerking motion and absorbing the shocks to a significant extent.

3.6.8   DRIVER SEAT

For Driver Seat, the seat of a bicycle is used. It is mounted to the Base Frame by welding another frame over the Base Frame.

3.6.9   PEDAL AND FRONT WHEEL

The pedal is mounted to the Front Wheel as well as the Base Frame by Two tubes. The front-wheel is taken from a ranger bicycle, as it has a larger surface contact area and is also suitable for carrying a heavy load.

3.6.10   COMPARISON OF EXISTING RICKSHAWS TO IMPROVED TRICYCLE

In Existing Rickshaws, the driving Sprocket has a 20 cm diameter, and the driven Sprocket has an 8 cm diameter.

Hence, $r_1 = 10$ cm and $r_2 = 4$ cm. Therefore, the ratio of $\left(\frac{r_2}{r_1}\right)$ becomes

$$\frac{r_2}{r_1} = \frac{4}{10}$$
$$=> \left(\frac{r_2}{r_1}\right) = 0.4 \qquad (3.33)$$

If there are four persons sitting in the passenger seat. Mass of each person = 70 kg.
Then, W = 70 × 9.8 × 4 = 2744 N

In the Improved Tricycle, the driving Sprocket has an 8 cm diameter, and the driven Sprocket has a 20 cm diameter.

Hence, $r_1' = 4$ cm and $r_2' = 10$ cm. Then the ratio becomes

$$\frac{r_2'}{r_1'} = \frac{10}{4}$$
$$=> \frac{r_2'}{r_1'} = 2.5 \qquad (3.34)$$

Therefore, the effort required for pedaling in the Improved Rickshaw is 2.5 times less than the existing Rickshaws.

The complete improved tricycle with four passengers with the driver/rickshaw puller is presented in Figure 3.27.

The cost estimation for the improved tricycle is given in Table 3.2 below

FIGURE 3.27 Developed tricycle with driver and passengers.

TABLE 3.2
Cost Analysis of Improved Tricycle Rickshaw

Sl. No.	The Name of Component	Price of (₹)	Units	Price (₹)
1	Rear Wheel	1250.00	2	2500.00
2	Front wheel including suspension	3000.00	1	3000.00
3	Rear wheel suspension	600.00	2	1200.00
4	Diameter of sprocket 86 mm	130.00	3	390.00
5	Big sprocket diameter 200 mm	150.00	6	900.00
6	Pipe	60.00/kg	15 kg	900.00
7	Bearing	80.00	6	480.00
8	Axle (Rear)	120.00	2	240.00
9	Seat (Rear)	500.00	2	1000.00
10	Seat (Driver)	500.00	1	1000.00
11	Seat mountings	1000.00	–	890.00
12	Frame and Hood	1500.00	1	1500.00
13	Shock absorbers	500.00	4	2000.00
Rupees sixteen thousand only				16000.00

3.7 CONCLUSIONS

The tricycle is manually propelled entirely and is significantly more efficient than traditional rickshaws. Because it is fully manual, there is no power or gasoline required to drive it. So, the alternative to electric or self-rickshaws is also entirely free of pollutants and cheaper. In addition, this trike is extremely convenient than any other rickshaw. In the future, this design will have a wide range. The following points can be concluded

- Conventional rickshaws utilize outdated technology, traditional design therefore, they are non-ergonomic in end uses.
- Attempts are made to develop a mechanically improved better quality and ergonomic tricycle rickshaw. The newly designed tricycle rickshaw has improved differential compared to the presently used one.
- Effort required for pedaling in Improved Rickshaw is 2.5 times less than the Existing Rickshaws.
- It has more than two commuter's carrying capacity. The three-wheeler will be fabricated using locally available standard materials from the market.
- For mathematical modeling and simulation of human-powered drive trains, simulation software MATLAB has been used.
- The thigh gives the maximum torque at nearly 6 seconds from the start of the stroke, while that for the shin is around 7.5 seconds.
- The tricycle was designed using CATIA V5 commercial software. The stress analysis of the frame and drive mechanism was performed with ANSYS.

- The minimum stress was 6.69×10^6 N/m^2 and a maximum of 7.71×10^7 N/m^2 at the rear part of the frame. The average 60 kg mass of passenger total traction load is 485.98 N at an average velocity of 1m/s and a minimum of 105.57 N without any passenger at a velocity of 2.25 m/s.
- Stress ranges from 1.75×10^8 N/m^2 at the right-hand part of the shaft to a maximum of 1.89×10^{10} N/m^2 at the left-hand part of the shaft.
- Tricycle is favourable for both the passengers and the driver.
- The Rickshaw may be used for other goods by certain modifications.

ACKNOWLEDGEMENT

Assam Science and Technology Environmental Council, Guwahati (Govt. of Assam) funded research project "Design and Development of an Improved Tricycle Rickshaw" 2021–2023.

REFERENCES

[1] N. Norcliffe, "Neoliberal mobility and its discontents of working tricycle in China's city," *Culture and Society*, vol. 2, pp. 235–242, 2011.
[2] http://www.bicyclehistory.net/bicyclehistory/hi story-of-tricycle [Accessed: 30/06/2020].
[3] M. Hickman, "A study of power assist bicycle rickshaw in India including fabrication and test apparatus," BS Mechanical Engineering Project Report. MIT, USA, 2001.
[4] F. Alam, P. Silva, and G. Zimmer, "Aerodynamic study of human powered vehicle," *Procedia Engineering*, vol. 34, pp. 9–14, 2012.
[5] D. Meyer, G. Kloss, and V. Senner, "What is slowing me down? Estimation of rolling resistances during cycling," *Procedia Engineering*, vol. 147, pp. 526–531, 2016.
[6] P. P. Dutta, D. Das, M. Dutta, A. K. Shukla, T. K. Gogoi, and A. Das, "Studies on green design and manufacture of hybrid vehicle," In: Proceeding of 5th International and 26th All India Manufacturing Technology Design and Research, 2014, U. S. Dixit Editor, 2014, IITG, Guwahati: India. pp. 565-1–565-6.
[7] P. P. Dutta, S. Sharma, A. Mahanta, A. Choudhury, S. Gupta, R. Gogoi, D. Baruah, K. Barman, T. K. Gogoi, and A. Das, "Development of an efficient hybrid tricycle," In: Proceeding of 5th International and 26th All India Manufacturing Technology Design and Research, 2014, U. S. Dixit Editor, 2014, IITG, Guwahati: India. pp. 572-1–572-7.
[8] P. P. Dutta, S. Sarmah, A. Mahanta, A. Choudhury, S. Gupta, R. Gogoi, and D. Baruah, (2014). "Design and development of an improved hybrid tricycle," In: Proceeding of International conference on Recent Advances in Mechanical Engineering *2014*, S. S. Gautam Editor. 2014, Excell India Pvt. Ltd, New Delhi: India.
[9] P. P. Dutta, M. Dutta, A. K. Shukla, S. S. Shukla, A. Konwar, V. Kumar, S. S. Sharma, and C. K. Bora, "Design FEA and fabrication of a hybrid all terrain trike," In: International Symposium on Aspects of Mechanical Engineering&Technology for Industries, *2014*, P. Lingfa, and S. S. Gautam Editors. 2014, Excell India Publisher, New Delhi, pp. 97–103.
[10] A. M. Paudel, and P. Kreutzmann, "Design and performance analysis of a hybrid solar tricycle for a sustainable local commute," *Renewable and Sustainable Review*, vol. 41, pp. 473–482, 2015.

[11] A. Dutta, and T. Jash, "Studies on energy consumption pattern in mechanized van rickshaws in West Bengal and the problems associated with these vehicle," *Energy Procedia*, vol. 54, pp. 111–115, 2014.

[12] P. P. Dutta, S. Sharma, N. Amin, H. Pegu, and R. Chutia, "Design and development of a CVT based transmission for a hybrid electric vehicle," In: Proceeding of International Congress of Renewable Energy: Renewable Growth Through Acedimia Industry Interface *2011*, S. K. Samadrshi, S. Mahapatra, and S. Paul Editors, 2011, Tezpur University, India. p.202–210.

[13] P. P. Dutta, A. Konwar, A. Baruah, and V. Sapat, "Design, fabrication analysis of a lawn mower. Sustainability, Inspiration, Innovation and Inclusion" (Proceeding of 5th annual international conference on sustainability - SUSCON V: 113-123: ISBN: 978-1-78635-414-3): Editors (T. K. Giri, S. Jaipuria, K. C. Das, A. Mukhopodhay, B. J. Gogoi, and M. Bhattacharya), Emerald, New Delhi, 2017.

[14] P. P. Dutta, N. Saikia, De, and S. Sharma, "Development of a low-cost dynamometer for performance testing of an electric vehicle," 57th Annual Technical Session of Assam Science Society, 2012.

[15] R. Bhukya, and SSSRS Duvvuri, "Solar electric bicycle using permanent magnet direct current motor: a realistic prototype," *International Journal of Engineering and Advanced Technology*, vol. 9, no. 3, pp. 2763–2767, 2019.

[16] R. S. Keote, P. Kale, C. Raut, B. Samavedula, R. Khawade, and A. Dumanwar, "Solar based smart lawn mower," *International Journal of Engineering and Advanced Technology*, vol. 9, no. 4, pp. 2190–2194, 2020.

[17] S. Neuroula, and S. Sharma, "Design and analysis of DYC and torque vectoring using multiple frequency control electronic differential in an independent rear wheel driven electric vehicle," *International Journal of Engineering and Advanced Technology*, vol. 9, no. 2, pp. 307–315, 2019.

[18] M. Pillai, N. Sharma, V. Bhandari, and R. Bansal, "Design of all-terrain vehicle gear box," *International Journal of Engineering and Advanced Technology*, vol. 9, no. 4, pp. 487–490, 2020.

[19] K. N. Rao, P. Ramesh, C. Ashokkumar, and A. Bannu, "Design and implementation of smart cart using labview," *International Journal of Engineering and Advanced Technology*, vol. 9, no. 2, pp. 946–949, 2019.

[20] R. C. de Resende, and L. M. Gonzaga, "Manual punch planter's design and development for small farmer," *International Journal of Innovation and Sustainable Development*, vol. 13, no. 1, pp. 79–97, 2019.

[21] R. S. Gautam, "Sensitivity analysis of stress distribution in bicycle frame," *Australian Journal of Mechanical-Engineering*. 10.1080/14484846.2020.1763546

[22] V. B. Bhandari, *Design of Machine Elemnts*. New Delhi: Tata McGraw Hill, 2013.

[23] R. Pergher, V. Bottega, and A. Molter, "Biomechanical model and control of human postural system and simulation based on state dependent riccati equation," *Mecanica Computational*, vol XXIX, pp. 6605–6618, 2010.

[24] R. R. Davis, and M. L. Hull, "Measurement of pedal loading in bicycling: I. Instrumentation," *J. Biomechanics*, vol. 14, no. 20, pp. 843–856, 1981.

[25] R. R. Davis, and M. L. Hull, "Measurement of pedal loading in bicycling: II. Analysis and Results," *J. Biomechanics*, vol. 14, 857–872, 1981.

[26] C. Hampali, and C. Bendigeri, Design, and development of solar electric tricycle: Material Today: Proceeding 2021: 10.1016/j.matpr.2021.01.527

4 Potential Bio Composites for Food Packaging
A Green and Sustainable Approach

Raghavendra Subramanya, Arjun S, Dheeraj KG,
Sarvepally S Gokul, Rishikesh M R, and
Amol S Pawar

Departmentof Mechanical Engineering, Sai Vidya Institute of
Technology, Bangalore, India

CONTENTS

DOI: 10.1201/9781003242291-4

4.1   INTRODUCTION

A keen interest in the implementation of innovative creative materials for a range of end-use applications has been prompted by growing environmental and ecological consciousness [1]. Food products are primarily protected from the elements and degradation by packaging, which also informs consumers about the contents and nutrition [2].

Food packaging has grown primarily to impart quality in food goods, as well as inform or provide knowledge to consumers about the qualities offered in the products. Restraint, shield, and communication are the three main functions of packaging.

The current regime in several parts of the world raises awareness about food safety due to numerous food-borne medical disorders. The bacterial pathogen is the agent that contributes to the public's knowledge of the problem, and a diverse range of food-producing organisms is the essential source to be noticed that results in illness prevention or treatment.

The most critical activity for the preservation and selling of food goods is packaging. Petroleum-based food packaging materials have been used for decades due to their non-biodegradability, which has serious negative consequences. The food packaging sector might undergo a big change as a result of biomaterials. It was demonstrated that compared to an active system that didn't reveal any agents, bioactive packaging material that produced bioactive substances was more effective. Materials for biodegradable food packaging made of biopolymers could be a good alternative to non-biodegradable plastics.

Because of their adaptability, durability, low density, ease of manufacture, low cost, and controlled wettability, polymers are utilised in food packaging. However, food quality problems might occur if it is packaged in polymers. Food wrapping is an essential component of the food business. since it helps with the secure, sanitary storage of food and beverages. On occasion, though, it can also raise questions about food safety. When heated, some packing materials, including Styrofoam, polythenes, and other types of plastic, can release compounds that are harmful to humans. Irradiated packaging materials have the potential to contaminate food with harmful substances that are not intended for consumption. Colours and packaging adhesives are only a couple of the many elements that go into food packaging.

The interaction of the outside ecosystem with the packaging food around the packaging material, as well as the adsorption/absorption of active ingredients on the product packaging, will have an influence on the packed food's quality and the integrity of the package [3]. Small-scale chemical migration from packing material has an impact on food sensory attributes, while large-scale migration may have detrimental health consequences. Penetration is the term for the transfer of volatile and aromatic elements, gases, moisture, as well as other low-molecular-weight molecules from the outside world into foodstuff or vice versa through the packaging material.

Quality deterioration may occur as a result of polymers absorbing aromatic food ingredients [4].

Interactions are influenced by the type of food material used in packing, particularly food-simulating chemicals such as moisture, oil, ethanol, hexane, and acetic acid [5].

Researchers are striving to create packaging materials that decompose biologically since synthetic polymer materials harm the environment. The creation of biodegradable polymers involves the utilisation of materials derived from plants, animals, and other renewable resources.

In addition to producing a lot of heat, making plastic also produces a lot of carbon dioxide. Paper has disadvantages when used to package food. Paperboard packaging is used to store and display food, pharmaceuticals, and hygiene items; corrugated containerboard is used to ship and carry goods including electronics, delicate glassware, and commodities. Customers have a sustainable choice for bringing their purchases home thanks to paper bags. Glass is thought of being a pure and secure kind of packaging. Be aware of the glass' brittleness and high processing costs if it is used to package goods. The production of common packaging materials requires a lot of energy, including plastics, corrugated boxes, and plastic bags. Burning is a common method of generating energy, and this process results in significant atmospheric emissions of greenhouse gases including carbon dioxide and methane. While this is happening, normal packaging waste contaminates the soil, water, and plants by being dumped in landfills or the ocean. Therefore, in order to stop all of these harmful impacts, we must utilise ecologically friendly packaging.

Bio-based packaging materials are getting more popular as their use expands. Bio-based packaging materials provide physical protection, convenience, and barrier protection. Polymers, paper (brick cartons, cardboard), foam, film, and other bio-based materials are currently available as alternatives to synthetic plastic, which is the most commonly used. This chapter provides a comprehensive understanding of the potential literature on processes and approaches for adjusting the performance and properties of bio-based materials to fully utilise them [6–8].

The range of composite research and material science topics is broad and cutting-edge. Modern scientific research is experiencing unprecedented discovery and evaluation. Packaging technology is an important industry in its own right, with its own particular point of view. The use of nanocomposites and composites in packaging technologies is an important part of this wider scientific endeavour. The true scientific and technical enlightenment of today will be guided by humanity's scientific intellect, the truth, and the need for energy and environmental sustainability.

4.2 BIODEGRADABLE POLYMERS

Some recent significant studies have revealed the inherent value of bio composites material processing and development, focusing on natural sources such as cellulose, starch, chitin, and chitosan; proteins such as casein and soy protein; and lignocellulose–based fibres from various resources, as well as their characteristics and potential applications [9–14].

Plastic has replaced paper as the most extensively used packaging medium for consumer goods in the textile and food industries. Plastic, however, is a significant cause for worry due to its durability and represents a serious hazard to the environment. Severe natural disasters have indeed been connected to wasted trash containers. Plastic manufacturing emits a great deal of heat as well as carbon dioxide. Paper-based food packaging has its own set of disadvantages.

Packages are shipped and handled using a corrugated carton board. Food, healthcare, and utilities are packaged on paper board for easy storage and display. Traditional packaging materials such as plastics, corrugated boxes, and plastic bags need a lot of energy to manufacture [15]. Burning fossil fuels releases a lot of greenhouse gases into the atmosphere, including carbon dioxide and methane, which are used to create energy, as opposed to traditional packaging, which is dumped in landfills or the ocean and contaminates the environment's soil, water, and vegetation. The adoption of ecologically friendly packaging is thus required by users to prevent all of these negative effects.

Goods made of biomass that is collected each year during agricultural harvests that are biodegradable Rich flour, soy products, cornstarch, waste cassava, green coconut husk extraction, sugarcane, and hemp all offer a foundation for environmentally friendly and sustainable solutions that might displace and eliminate petroleum-based composites and polymers [16].

By promoting the use of recyclable materials as well as low-impact production techniques, bio-based products protect resources. Furthermore, bio-based products reduce exposure to potentially hazardous and harmful substances. Eco-friendly and compostable materials such as cloth, paper, and jute handbags are available at stores for carrying things.

The unplanned use of synthetic polymers is causing an increase in solid waste in the natural environment. This has an impact on the natural system and creates a variety of environmental hazards. Plastics are viewed as a threat to the environment because they are difficult to degrade.

To replace polymers in the environment, a global research effort is being made to create biodegradable polymers for waste management. The chemical breakdown of materials by microorganisms, including bacteria, fungi, and algae, is known as biodegradation. In the most typical scenario where they are disposed of, biodegradable materials have been demonstrated to disintegrate within a year into non-toxic carbon-containing soil, water, or carbon dioxide by natural biological processes [17].

4.2.1 BIO-BASED POLYMERS USED FOR FOOD PACKAGING

Due to their biodegradability, biocompatibility, and physical and chemical characteristics that are equal to those of conventional and non-biodegradable polymers, several polymeric materials, such as aliphatic polyesters, have generated a great deal of attention. These polymers are employed in many different applications, including crop irrigation, packaging, and films.

The development of delivery methods for therapeutic substances such as hormone-disrupting steroids, anticancer drugs, antimalarial drugs, and peptide hormones has also made extensive use of polylactic acid (PLA). With the necessary modifications, PLA has been utilised to make a range of drug-release systems, including microspheres, rods, films, and nanoparticles. The inclusion of lactide monomers in PLA is a noteworthy benefit. The molecular weight can also alter when functional groups are added to the foundation. These functional groups can also include hydroxyl species, lactic acid, and water. To obtain the desired qualities, the user can alter the PLA polymer's backbone configurations.

L-lactic acid and D-lactic acid are the two chiral variants of PLA that are present. Stereoisomers allow for the differentiation of PLA types. The polymer chain in PLA is hydrolyzed during the process of degradation. The hydrophobic properties of PLA might potentially trigger an inflammatory reaction in the surrounding tissue [18].

The most basic linear aliphatic polyester is polyglycolide, often known as polyglycolic acid (PGA). It is a naturally occurring compound that is made from glycolic acid by ring-opening polymerisation or polymeric condensation. Absorbable sutures are frequently created using polyglycolide and its co-polymers with lactic acid, e-caprolactone, and poly trimethylene carbonate.

PHAs, or poly hydroxyl alkanoates, are biopolymers having a wide variety of characteristics, including rigidity and high crystallinity as well as flexibility, amorphousness, and elastomeric qualities. due to both these qualities and their innate biodegradability. As a result, these biomaterials have a lot of promise for use in medical implants. When the fungus is common, the synthetic biodegradable polymer polyvinyl alcohol is likely to be subjected to biodegradation. Due to random endo breakage of the polymer chains in an aquatic environment, PVA has high water solubility. Biopolyesters are organic, bacterial-derived macromolecules. Despite their expensive price, they are becoming more and more popular. It results from the way they are produced, utilised, and disposed of, which is ecologically beneficial. As cellular and energy storage materials, polyhydroxyalkanoates (PHAs) are synthesised. They typically form as a result of nutrient deficiency. PHAs are completely biodegradable and have similar properties to traditional plastics.

Polyesters, a type of polymer with a functional group of ester, come in both thermoplastic and thermoset varieties. However, thermoplastics make up the bulk of polyesters. Biodegradable polyesters include some synthetic and natural varieties. Biodegradable natural polyesters contain naturally occurring compounds like plant cuticle chitin [19].

4.2.2 Natural Polysaccharides

With a chain length of 500–2,000 glucose units and 1,4 glycosidic connections connecting its repeating glucose groups, starch is a straight or cross-linked polysaccharide. Both amylase and amylopectin are essential components of starch. For a range of uses, including food, paper, textiles, and biological applications, chemically altered starches have been developed. To produce premium packaging sheets and films, aliphatic polyesters and biodegradable starch are mixed. Compost films, bread pouches, overwraps, pieces for "flushable" sanitary goods, and shopping bags are a few examples of commonly accessible commercial products constructed of starch.

If starch isn't initially treated by plasticisation, combining with other materials, genetic or chemical alteration, or a combination of the aforementioned, it won't form films with the necessary mechanical attributes, such as physical and mechanical properties. To reduce film brittleness and increase flexibility and extensibility, compacting agents are frequently used. These agents include fatty acids, lipids, and compounds, as well as glycerol, sorbitol, and polyethylene glycol.

Antimicrobial packaging, a type of active packaging, stops or slows the growth of bacteria that are present in packed food or packaging materials. Food goods frequently contain preservatives such as organic acids, alcohol, antimycotics, enzymes, and oxygen absorbers [20].

4.2.3 Cellulose

The most common organic compound is cellulose. It is a major element of green plants and algae's main cell walls. Wood pulp and cotton are the most common sources of cellulose. It's mostly used to make cardboard and paper.

Fresh fruit and vegetables have been protected using composite films composed of CMC (carboxymethyl cellulose), which have a higher moisture absorption and permeability and may also absorb the water produced by the fresh produce during respiration to stop food spoilage. Utilising a bio-derived composite film for food preservation may enhance the product's quality and safety. Despite the film's great moisture permeability, the pace at which moisture is absorbed depends on the amount of CMC. This characteristic is connected to the hydroxyl group on the surface of CMC, which is readily absorbed by water molecules and makes the CMC composite film hygroscopic [21].

In addition to jellies, toothpaste, diet pills, water-based paints, detergents, textiles, and a range of paper goods, many wheat-based items also include cellulose [22].

Casein and soy protein are made into soy protein polymers, which have the potential to be biologically active. About 20% of soybeans are made up of oils, while 40% are made up of proteins. In certain species, the protein content can reach up to a whopping 55%. Soy proteins have adhesive properties and are employed in a variety of items, including automotive parts. Dried soy polymers have a 50% greater modulus than epoxy engineering plastics [23].

4.2.4 PROCESS OF BIODEGRADATION

Degradation, in general, is the process through which a polymer's properties deteriorate due to a variety of stimuli, such as light, heat, UV radiation, mechanical stress, and so on. The degradation causes the item to become brittle, the lifespan of the material is reduced, and the tiny pieces that develop don't really add to the mechanical qualities as efficiently.

While living microorganisms like bacteria, fungi, and algae cause bioplastics to degrade, biodegradability refers to a material's ability to be safely broken down by microbes into carbon dioxide, biomass, and water and used as a source of carbon.

4.2.5 PROPERTIES OF BIO-BASED MATERIALS

Better bio-based product characteristics and performance are often defined by efficiency, efficacy, cleanliness, and recycling. By efficiently enclosing and safe-guarding items—two crucial aspects of packaging—bio-based materials help promote green production. Additionally, the packing technique is made to use energy and materials efficiently, producing less waste.

Features including gas and water vapour permeability, transparency, printability, accessibility, UV resistance, water resistance, as well as acid, alkali, and grease resistance, are all significant factors to consider when it comes to food bio-based materials.

4.2.6 WATER VAPOUR TRANSMITTANCE, GAS BARRIER, AND MOISTURE RESISTANCE

The main components of the gas mixture inside the package are carbon dioxide, oxygen, and nitrogen. Mineral oil-based polymers are used in bio-based goods with gas barrier properties. A highly constrictive water vapour barrier is created when moisture-sensitive PA6 (polyamide) or EVOH (ethylene vinyl alcohol) are mixed with LDPE, keeping moisture out of the food goods [24].

4.2.7 PHYSICAL BLENDING AND CHEMICAL PROCESSING

Bio-based composites are frequently utilised as structural components in textiles, packaging, and disposable items, but they may also be employed in the building and transportation sectors to reduce weight, increase cost-effectiveness, and improve load-bearing capacity. If qualities, processing, and therefore performance is enhanced, this can be done [25].

Specific mechanical characteristics of packing materials are a need for maintaining food safety and determining the length of the guarantee term. Mechanical and thermal qualities are essential in the creation and usage of goods generated from materials [26].

The mechanical and barrier characteristics of egg-albumen films were studied by Gennadios et al. Aqueous egg albumen solutions with various concentrations of GLY, PEG, sorbitol (S), and plasticisers are used to create egg albumen movies.

The researchers concluded that PEG plasticised films own a higher tensile strength than GLY plasticised films [27].

The mechanical properties of foam have a significant link with its density, and this property is typically the most important. Other aspects of particle foams, such as the foam's molecular structure or the extent of particle bonding, have an impact on the foam's final attributes. PLA particle foams are a novel material class [28].

4.3 APPLICATION OF BIO COMPOSITES IN THE FIELD OF FOOD & PACKAGING INDUSTRY

Future food and packaging sector workers should help lessen the waste of food and packaging materials and its detrimental effects on the environment. Plastic has been a primary material used in the packaging industry's utensils, bags, packages, etc. The biggest use of plastics is in packaging, which accounts for roughly 43% of all users in our nation and 35% worldwide (till 2013) [29].

In the food and packaging industries, starch and poly lactic acids (PLAs) are the most often utilised materials; additional bio-materials including gluten, zeins, and prolamine derived from corn are also employed. Due to their commercial availability, high processability, rigidity, impact and tensile strength, barrier qualities, and balance of factors affecting degradation, PLAs are the most intriguing materials for food packaging. PLA was reinforced with hemp and was injection moulded to obtain composite material where silver nano-particles were introduced into hemp to achieve antimicrobial properties. Chitosan and Zein have some interesting characteristics; Chitosan and antimicrobial properties and water resistance of zein impart large importance in developing packaging materials [30,31].

4.3.1 APPLICATION OF NANOCLAY REINFORCEMENT IN FOOD PACKAGING MATERIALS

To improve consumer safety and health. The main goal of the active components feature was to increase the shelf life of a product by incorporating parts that would allow chemicals to permeate through the packaging of the food or into the surrounding environment or drain into it. The development of both active and intelligent product packaging has been significantly influenced by the evolution of nanotechnology.

When combined with bovine gelatin polymer, halloysite nanoclay showed stronger mechanical capabilities than nanosilica, as well as greater barrier characteristics and water solubility, according to Voon et al. Nanoclays are perfect for use as a reinforcement agent in packaging or containers of fast-moving consumer goods, such as food and drinks, due to their low cost [32]. Additionally, nanoclays are not too pricey. The addition of a little quantity of nanoclay can significantly enhance the host polymer's barrier, mechanical, thermal, and degrading characteristics. (10 wt per cent) [33,34].

Nan clays are distinguished from other varieties of clay by their distinct particle shape, flake-soft content, low specific gravity, and high aspect ratio with nanoscale thickness. To increase the properties of the polymers, several kinds of nanoclays are added [35]. Owing to its unique surface area with a comparatively large aspect ratio (50–1000) and good compatibility with many organic thermoplastics, montmorillonite (MMT, MMT-Na+) and organophilic MMT (organic modified MMT, OMMT) have attracted the most interest from industry and academic researchers in the packaging field [36].

The incorporation of nano clays into organic polymers is influenced by a number of variables, such as the type of polymer and nano clay, loading quantities, manufacturing methods, desired features and uses, and side effects (i.e., colour alteration, change in elongation, or surface roughness). Varying surface enhancers are used to create organically modified MMTs with different characteristics. Due to issues with trash disposal, packaging for fast-moving consumer items, particularly food and beverages, has been a major environmental concern worldwide.

Different polysaccharide and protein films are stiffer and less durable than thermoplastic starch films. Because polysaccharides are hydrophilic, starch films have excellent water vapour permeability but low mechanical characteristics relative to commercial polymers. To compensate for these drawbacks, starch-nanoclay hybrids have been developed. With the addition of unaltered hydrophilic nano clay, significant increases in the barrier and mechanical characteristics were observed [37–39]. The hydrogen bonding interaction has the potential to enhance the dissemination of nanoclay throughout the starch matrix since free hydroxyl groups are readily available in starch.

Intelligent Packaging Colorimetric Indicator System. Nanoclay was used in a sophisticated and perhaps practical way by Gutierrez et al. as a colourimetric indicator for smart packaging. By injecting a blueberry extract between the nanoclay particle interlayers, the indication system was created. Anthocyanins found in MMT and OMMT powder responded well to pH changes in both acidic and alkaline environments [40].

Biodegradability Enhancement. Food is the single item that every person eats three times every day. As a result, more than two-thirds of all packaging waste is accounted for by food packaging [41]. Common treatment and disposal methods for unused packaging include composting, deterioration, landfilling, burning, and recycling. Deterioration in the ecosystem is commonly preferred due to the shorter degradation time and reduced energy need. Recent studies have demonstrated that the inclusion of OMMTs considerably increases the biodegradability of polymers derived from both petroleum and biological sources through a number of breakdown mechanisms [42,43]. Petroleum-based polymers can experience abiotic degradation due to changes in their physical and chemical properties when exposed to heat or photo radiation. It has been found that by increasing the Transmittance capabilities of polymer nanocomposites, nanoclays reduce the abiotic/oxidative degradation of polyolefins [44].

According to the formation of carbonyl and hydroxyl species during oxidation, Kumanayaka et al. found that the photo-oxidation of LDPE/OMMT was larger than that of virgin polyethylene. The breakdown of alkyl ammonium ions in the OMMT, which led to the creation of tertiary amine, olefin, and acidic sites on the clay particles, sped up this process [45].

Due to their well-established functional characteristics of reinforcement and barrier strengthening, nanoclays are increasingly being used in polymer-based packaging solutions. While thermoplastic starch's mechanical and barrier qualities, as well as the biodegradability of synthetic polymers, are all improved by unaltered nanoclays, their inherent hydrophilicity hinders their inclusion with organic polymers. Organophilic nano clays are used not only to increase polymer dispersion but also to improve the qualities of packaging materials. Many studies have shown that nano clays might be used as an antibacterial agent, to regulate and discharge active chemicals, as a colourimetric indicator template, and as a colourimetric indicator template in food packaging, expanding the range of applicability of nano clays.

4.4 APPLICATION OF BIO COMPOSITES IN THE FIELD OF MEDICAL INDUSTRY

Substituting materials in medical applications is a complex task and it is a challenge in current times to accomplish maximum efficacy without revision surgeries. The bio-composite used in medical applications should be corrosion and wear-resistant. Human nature will be out of the blue and hence the bio-composite materials should be strong and flexible enough to accommodate various kinds of motions that a human undergoes in his daily life. The emergence of a new generation of hybrid nanostructured materials has paved the way to make biomaterials and apply them in medical applications [46].

The major reason for adapting biomaterials in artificial prosthetics:

- It degrades inside the body which eliminates the surgery or operations to remove prosthetics after its failure.
- Less harmful compared to metals and ceramics/plastics.

Surface finishing is one of the important factors in making implants that significantly increase longevity or durability with minimal failures. The finish of the implant should be smoothly polished, rough blasted surface or geometrically textured surfaces. In Hip Joints prosthetics, biopolymers are used as a matrix with reinforced materials like woven fibres or nanofillers. Biomaterials are used in bone plates, surgical instruments, skin repair tissues, dental implants, contact lenses, bone cement, medical laboratory apparatus, stents, heart valves, knee replacements, cardiac pacemakers, etc., [47].

4.5 MANUFACTURING METHODS OF FOOD PACKAGING MATERIALS

Recently, different techniques for addressing the issues in food industries for the creation of packaging materials have attracted attention. The extrusion process is given special attention since it rules the critical industrial procedures for the creation of films, sheets, trays, and items. The section also briefly discusses foamed materials, and it closes with a discussion of finishing packing procedures including coating and lamination.

4.5.1 EXTRUSION

The extruder, a huge mixer with a continual conical screw, is at the core of the bulk of industrial operations used to generate packaging materials. A hopper feeds resin grains into the extruder, which then transports the granules to the screw's neck and compression zone. The thread crushes into the resin space as a result of the screw's spinning motion, causing volumetric compression and a significant increase in the shear forces.

As a result of the massive energy input, the substance melts (pressure, friction, and shearing forces). The development of a homogeneous fluid is aided by direct contact with Heating systems placed along the casing length, from which the molten material is pushed. A slot, sometimes referred to as a die, is located at the extruder's end and controls the final shape, called an extrudate [48]. A variety of operating parameters must be altered based on the kind of polymer being processed to pump the material, which influences the polymer's resistance to movement against the nozzle. To get the optimal extrusion performance, the number of screws, as well as their design, positioning, and geometry, must be carefully determined. While twin screw extrusion is becoming common because of advantages including higher production uniformity, increased efficiency, and increased flexibility, a single spinning screw is still used in some circumstances (for instance, in a basic polymer framework).

Contrary to single-screw extrusion, twin-screw extrusion consists of two screws and allows for the mixing of a larger variety of polymers, including bio-based polymers. The majority of polymers used in food packaging, including nylon, polypropylene (PP), polystyrene, and polyethylene terephthalate (PET), are manufactured by extrusion (PS). According to the variances in these polymers' characteristics, industrial requirements, and desired end material form, several extrusion processes can be applied, as explained in the following sections. Up to execution, every one of these processes is identical.

4.5.2 BLOWING

The process of filling a molten tube with air or nitrogen as it exits the nozzle is known as blowing. Blown film frequently rises vertically. The most common and feasible nozzle form is round (annular), but the product quality decreases as a result. Spinnerets, particularly spiral nozzles, transcend annular nozzle restrictions; nonetheless, they are more sophisticated solutions [48]. After that, the small bubble is flattened, coiled around a reel, and passed through a succession of nip rollers. The copolymer links are oriented in the right direction by pulling and blasting.

The finished film will be heavier in the rolling direction depending on whether it was pulled or blasted. The blow-up ratio is the ratio of the diameter of the bubble to the diameter of the nozzle.

4.5.3 CASTING

The tube that emerges from the extruder, as well as the T- and coat hanger-shaped die, are used to create films and sheets. These geometries enable the transformation of a spherical melt into a flow that resembles a flat and level surface. The solid melt is quickly cooled as it leaves the die using quenching. The film is fastened to the cold roll using a sophisticated electrostatic device. An important part of the production in cast film is the draw-off ratio, which would be the relationship between draw-off speed and extrusion speed [48]. The cast film process is more costly and less adaptable than the blowing method.

4.5.4   BLOW MOULDING

The most common product for the food packaging industry is bottles, which could be produced using the blow moulding technique to generate three-dimensional hollow bodies. The fundamental idea behind blow moulding is to fill a thermoplastic pipe that is encased in a cooled mould with air or nitrogen to give the product its precise final shape. Depending on whether the polymer solution is formed in the same equipment in which the blowing occurs, there are two main blow moulding processes: (1) extrusion blow moulding, where the polymer solution is formed in the same equipment where the blowing occurs, and (2) injection blow moulding, where the polymer solution is moulded individually.

4.5.5   EXTRUSION BLOW MOULDING

In the steady process of blow moulding, the homogenous mixture is extruded into a mould that has the extruder's die. When the parison is the proper size, the open mould halves are covered, and an air stream is fired from a pin inserted within the die head, allowing the parison to expand until the desired mould size is reached. When the item has cooled, the mould is opened and it is taken out [49]. When making bottles and containers, extrusion blow moulding is a flexible technique, especially when employing HDPE and PP polymers.

4.5.6   INJECTION BLOW MOULDING

In this type, the component parison is originally made in a mould that is connected to the back of the extruder. While the parison is still partially molten, it is transferred to a second mould, where pressurised air is injected into it until it takes on the correct shape and size. Injection blow moulding is therefore a continual process. The mould is opened and the piece is thrown out once it has cooled [49]. Because of the functional qualities of the bi-exceptional orientation, notably the production of an obstruction to gases and vapours, this approach is particularly significant for PET bottles meant for drinks containing dissolved gases such as carbon dioxide.

4.5.7   INJECTION MOULDING

For the continuous production of three-dimensional items with cavities, such as coffee bean powder capsules and yoghurt jars made of PP, injection moulding is used. Using an extruder to melt the polymer, the melt is then forced into a mould that has been designed to suit the extruder's die. The object may quickly cool down because a precise quantity of the soft polymer makes it through the mould's cold walls [50]. One of the two halves of the mould separates from the other after cooling, allowing the item to be released from the mould.

4.5.8 THERMOFORMING

The most common technique for making film sheets, containers, and mugs from a moderately thick sheet—typically between 100 and 300 mm—is thermoforming. The plastic sheet softens when heated and may be stretched to take on its final shape. Vacuum or mechanical compression is used to push the material against the profile of the mould during this process. The object is emptied and given its final form after cooling. Thermoforming is a thinning procedure that affects the starting sheet by stretching it. This factor needs to be closely watched when working with multiple-layer materials [51].

4.5.9 COMPRESSION MOULDING

Compression moulding is a technique for imparting mechanical stress to a polymeric material that has been placed in the bottom of a mould. Conduction causes the substance to reach its melting point. The mould's top is then forced down until it completely coincides with its opposite. The polymer melt flows into the gap between the two tool components until it is filled by pressure and heat. This process could involve curing or cross-linking, which gives the finished product mechanical resistance, insolubility, or thermal stability. Once the object has cooled, the mould may be opened to release it. Screw thread caps for bottles and other fluid containers are created using this technique.

4.5.10 FOAMED PLASTICS

Foamed plastics, sometimes referred to as organic plastics or plastic foams, are a growing subcategory of materials used to package food because of their unique characteristics, which make them the best option for certain uses [52]. For instance, foams are widely renowned for their ability to cushion, are lightweight, and have a fair amount of strength such that they are suitable for use in perishable meat and fish packing. Although practically any plastic polymer may be used to create foams, the most often used ones in food packaging are polyurethane (PU) and continued-expansion polystyrene foams, for situations requiring direct food exposure, both are favoured above polyurethane foams. Both physical and chemical agents might produce pores during the foam formation process.

4.5.11 CO-EXTRUSION AND LAMINATION

The majority of foods require packaging to maintain quality and safety throughout their shelf life. To achieve this, several food packaging materials must be multifunctional.

This cannot be done with a single polymer, thus mixing polymers with different properties is an efficient way to create meaningful packages [53]. Lamination and co-extrusion are the two fundamental processes used to create multi-layer materials. Co-extrusion is a one-step procedure for creating composite films from a range of materials by integrating a single die head with two or more separate extruders. The capacity to mix thermoplastics with non-plastic components like metals or cellulosic materials is the fundamental advantage of coextrusion lamination. For instance, lamination is used to create a mixture of PE, cardboard, and aluminium foil.

4.5.12 Coating

Even without adhesives, as in lamination, Once an item is coated, at minimum one layer of a liquid or molten substance is spread to its surface. Since coating technology may be applied to almost any kind of film, sheet, or structured structure, it offers a complete answer (such as bottles and jars). Stretchy packing sheets are typically coated using one of four methods: rolling, dipping, spraying, or brushing. The impact of plastics discharged into the environment and the problem of open dumping have both gotten a lot of attention recently. In order to decrease the number of petroleum-derived goods required upstream, coating technology has drawn more attention.

According to the package optimisation theory [54], such coatings enable the production of lightweight plastic wrapping while keeping certain material features including heat conditions, surface characteristics, gas and vapour insulation characteristics, and optical and interface qualities.

4.5.13 Liquid Coating

This method makes use of liquid solvent-free paints as well as paints made with aqueous or organic solvents. Aqueous or organic solvents are used to dissolve or scatter the solids in the beginning. The process's solvent concentration and solvent amount relative to the solid are crucial because they affect several crucial aspects, including the system's rheological properties and the solution dispersion [55].

4.5.14 Extrusion Coating

The name of the technology suggests that molten thermoplastic material is imposed as soon as it leaves the extruder die. Despite being straightforward, this method is still widely used in the food packaging sector, particularly when LDPE is deposited as the sealing layer for cellulose goods. Unlike lamination, extrusion coating permits the deposition of very light layers of different materials, which requires large prepared sheets [56,57]. Extrusion coating also helps manufacturers create unusually high binding forces between the polymeric and the cellulose base by allowing the melt to permeate into the porous paper substrate. For extrusion coating, ionomers are used in addition to LDPE, PP, and PA.

4.5.15 METALLISATION

The process known as "metallisation" uses heat evaporation to create a very thin layer of metal on a solid matrix. A small, high-vacuum chamber is used to heat wires of a solid metal, often aluminium, to high temperatures in the "metallisation" physical vapour deposition process. In these conditions, the metal sublimates, releasing extremely minute particles that settle on the substrate in all directions. In reality, an inappropriate or erroneous operation can result in design attributes and arrangements that do not fulfil the technical and operational criteria for effective encapsulation, protection, and food shelf life enhancement. Furthermore, with the introduction of novel materials like bioplastics, nanocomposites, etc., customised processes and customised solutions are important to expanding into new markets.

4.6 MISCELLANEOUS

Considering the well-known challenges of petroleum scarcity and growing interest in reducing the environmental burden caused by the widespread usage of petrochemically derived polymers, the development and use of biodegradable starch-based materials have received increased attention.

Nutritional enrichment using phenolic acid-rich foods has been found to give antitumor, anti glycemic, and antioxidative effects, which may be used to make healthy foods. Grain cereal cereals are increasingly being used in food formulations across the world since they are high in phytochemicals and dietary fibre, both of which have multiple health advantages. Millets are significant crops in semiarid and tropical parts of the world because of their pest and disease resistance, short growing season, and production in tough and drought circumstances when other cereals cannot be depended on to deliver sustainable yields.

The author developed a biodegradable spoon out of millet and starch. Initially, grinders are used to crush millets into powder form. This millet powder was combined with starch and water. The mixtures were transferred to a steel mould with a core and cavity shaped like a cutlery spoon (shown in Figure 4.1a). The mould was held under compression using a universal testing machine (Figure 4.1b and c) and heat was applied through heating coils fixed under the mould to a temperature of $110°C$ (8 bar pressure is maintained for 15 min) The spoon was ejected from the mould (Figure 4.1d).

These millet bio cutlery spoons, like ice cream wafers, will be used to devour meals and then consumed. There are no disposal difficulties, and a synthetic spoon may be conveniently replaced with this. More than ever, millets are a healthy grain that may be enjoyed by everybody. Small tweaks or improvements in materials, for example, will aid in increasing the health of the ecosystem.

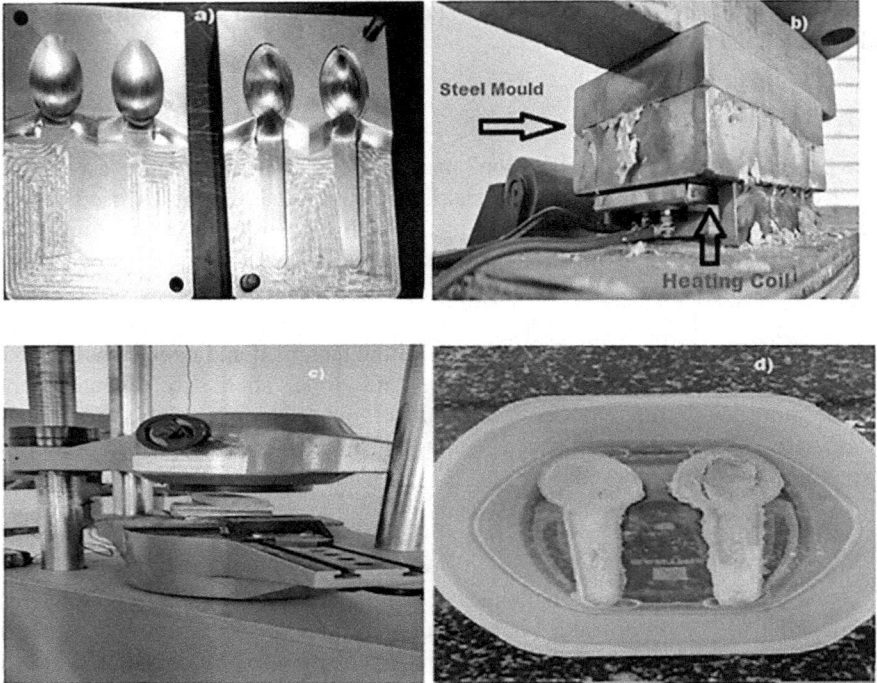

FIGURE 4.1 a) Steel mould having cutlery spoon shape core and cavity. b) Heating coils fitted under steel mould. c) Mould kept in a universal testing machine for compression load. d) final product cutlery spoon removed from the mould.

4.7 CONCLUSION

The chapter discussed vital variables to consider while utilising biobased polymers, which are employed in several industries such as food packaging, medical applications, and so on.

Bio-based materials are remarkable because they minimise a product's environmental impact throughout every phase of its life cycle. The planet's resources are being greatly impacted by the world's population rise and the expansion of the global economy. The food, textile, and plastic packaging industries have risen to the top of the sustainability agenda despite not being a large contributor to environmental issues due to their substantial human effect. Recycling bio-based materials will be much more ecologically benign than composting as a last option as bio-based goods gain popularity.

It is anticipated that a thorough knowledge and use of these emphasised elements would result in system-based solutions on a global market for the effective deployment of active manufacturing procedures.

To sum up, these are the most important takeaways from this chapter:

- Synthetic plastics will be replaced with biodegradable materials.

- Bone plates, surgical instruments, skin repair tissues, dental implants, contact lenses, bone cement, medical laboratory gear, stents, heart valves, knee replacements, cardiac pacemakers, and other biomaterials are all employed in medical devices.
- Green Biocomposites have a wide range of qualities that can be employed in a variety of applications. The most commonly utilised ingredients in the food and packaging industries are starch and PLA (Poly Lactic Acid).

ACKNOWLEDGEMENT

The authors wish to place on record their sincere thanks to Vision Group of Technology, Govt. of Karnataka for providing funds (DST/KSTePS/VGST/2020-21:GRD-990) for carrying out the investigation. Also, Sai Vidya Institute of Technology, Bangalore for their support.

REFERENCES

[1] A. S. Singh, and V. K. Thakur, "Mechanical, Morphological, and Thermal Characterization of Compression–Molded Polymer Biocomposites," *Int. J. Polym. Anal.Charact*, vol. 15, pp. 87, 2010.

[2] R. Coles, D. McDowell, and M. J. Kirwan 2003. "Introduction," In: Coles R., McDowell D., Kirwan M. J. (eds), *Food Pack Technol*. Blackwell Publishing, London, pp. 1–31.

[3] S. G. Gilbert, J. Miltz, and J. R. Giacin, "Transport considerations of potential migrants from food packaging materials," *J Food Process Preserv*, vol. 4, pp. 27–49, 1980.

[4] M. G. Sajilata, K. Savitha, R. S. Singhal, and V. R. Kanetkar, "Scalping of flavors in packaged foods," *Compr Rev Food Sci Food Saf*, vol. 6, pp. 17, 2007.

[5] I. S. Arvanitoyannis, and K. V. Kotsanopoulos, "Migration phenomenon in food packaging. Food package interactions, mechanisms, types of migrants, testing and relative legislation-a review," *Food Bioprocess Technol*, vol. 7, pp. 21–36, 2014.

[6] B. Abbes, F. Abbes, and Y-Q Guo, 2015. "Interaction phenomena between packaging and product," In: S. Alavi, S. Thomas, K.Sandeep et al (eds), *Polymers for packaging applications*, 1st edn. Apple Academic Press, New Jersey, pp. 39–70

[7] L. Castle, A. Mayo, C. Crews, and J. Gilbert, "Migration of poly (ethylene tereph-thalate) (PET)oligomers from PET plastics into foods during microwave and conventional cooking and into bottled beverages," *J Food Prot*, vol. 52, pp. 337–342, 1989.

[8] M. Duran, M. S. Aday, N. N. D. Zorba et al., "Potential of antimicrobial active packaging "containing natamycin, nisin, pomegranate and grape seed extract in chitosan coating" to extend shelf life of fresh strawberry," *Food Bioprod Process*, vol. 98, pp. 354–363, 2016. 10.1016/j.fbp.2016.01.007.

[9] T. Gurunathan, S. Mohanty, and S. K. Nayak, "A review of the recent developments in biocomposites based on natural fibres and their application perspectives," *Composites: Part A*, vol. 77, p. 1, 2015.

[10] A. Pappu, V. Patil, S. Jain, A. Mahindrakar, R. Haque, and V. K. Thakur, "Advances in industrial prospective of cellulosic macromolecules enriched banana biofibre resources: A review," *Int. J. Biol. Macromol.*, vol. 79, pp. 449, 2015.

[11] A. S. Singha, and V. K. Thakur, "Synthesis and Characterization of Short GrewiaoptivaFiber–Based Polymer Composites," *Polym.Compos.*, vol. 31, pp. 459, 2010.

[12] V. K. Thakur, and M. K. Thakur, "Recent advances in green hydrogels from lignin: A review," *Int. J. Biol. Macromol.*, vol. 72, 834, 2015.

[13] J. K. Pandey, S. H. Ahn, C. S. Lee, A. K. Mohanty, and M. Misra, "Recent advances in the application of natural fiber based composites," *Macromol.Mater. Eng.*, vol. 295, pp. 975, 2010.

[14] F. P. La Mantia, and M. Morreale, "Green composites: A brief review," *Composites: Part A*, vol. 42, pp. 579, 2011.

[15] G. K. Deshwal, N. R. Panjagari, and T. Alam, "An overview of paper and paper based food packaging materials: health safety and environmental concerns," *Journal of food science and technology*, vol. 56, no. 10, pp. 4391–4403, 2019.

[16] K. Dybka-Stępień, H. Antolak, M. Kmiotek, D. Piechota, and A. Koziróg, "Disposable Food Packaging and Serving Materials—Trends and Biodegradability," *Polymers*, vol. 13, p. 3606, 2021. 10.3390/polym13203606].

[17] R. Gautam, A. S. Bassi, and E. K. Yanful, "A Review of Biodegradation of Synthetic Plastic and Foams," *ApplBiochemBiotechnol.* p. 141, 2007.

[18] V. DeStefano, S. Khan, and A. Tabada, "Applications of PLA in modern medicine, Engineered" *Regeneration*, vol. 1, pp. 76–87, 2020.

[19] K. Leja, and G. Lewandowicz, "Polymers Biodegradation and Biodegradable Polymers – a Review," *Polish J. of Environ. Stud*, vol. 19, no. 2, pp. 255–266, 2010.

[20] N. Khairuddin, and I. Muhamad. "Antimicrobial Effects on Starch-Based Films Incorporated with Lysozymes," *Pertanika Journal of Sciences and Technology*, vol. 17, pp. 1–8, 2009.

[21] S. Bajpai, N. Chand, and S. Ahuja, "Investigation of curcumin release from chitosan/cellulose micro crystals (CMC) antimicrobial films," *Int. J. Biol. Macromol.*, vol. 79, pp. 440–448, 2015.

[22] D. Lavanya, P. K. Kulkarni, M. Dixit, P. K. Raavi, and L. NagaVamsi Krishna, "Sources of Cellulose and Their Applications –A Review," *IJDFR*, vol. 2 Issue 6, Nov.-Dec. 2011.

[23] M. Arif, L. S. Chia, and K. P. Pauls, "Protein-Based Bioproducts," *Plant Bioproducts*, pp. 143–175, 2018. 10.1007/978-1-4939-8616-3_9.

[24] S. Fischer, Vlieger de J Kock T., J. Gilberts, et al., "Green composites—the materials of the future a combination of natural polymers and inorganic particles," In: Proceedings of the Food Biopack Conference, p 29, 2000.

[25] J. Jefferson Andrew, and H. N. Dhakal, "Sustainable bio-based composites for advanced applications: recent trends and future opportunities – A critical review," *Composites Part C: Open Access*, vol. 7, pp. 100220, 2022.

[26] M. Iguchi, S. Yamanaka, and A. Budhiono, "Bacterial cellulose—a masterpiece of nature's arts," *Journal of Materials Science*, vol. 35, pp. 261–270, 2000.

[27] A. Gennadios, C. Weller, M. Hanna, and G. Froning, "Mechanical and Barrier Properties of Egg Albumen Films," *Journal of Food Science*, vol. 61, pp. 585–589, 1996. 10.1111/j.1365-2621.1996.tb13164.x.

[28] N. J. Mills, 2007. "Product packaging case study," In: N. J. Mills (ed) *Polymer Foams Handbook*. Butterworth-Heinemann, pp. 281–306.

[29] V. Guillard, S. Gaucel, C. Fornaciari, H. Angellier-Coussy, P. Buche, and N. Gontard, "The Next Generation of Sustainable Food Packaging to Preserve Our Environment in a Circular Economy Context," *Frontiers in Nutrition*, vol 5, 2018.

[30] H. N. Salwa, S. M. Sapuan, M. T. Mastura, and M. Y. M. Zuhri, "Green Bio composites For Food Packaging," *International Journal of Recent Technology and Engineering (IJRTE)*, vol 8, pp. 450–460, 2019.

[31] A. Jiménez, and R. A. Ruseckaite, 2012. "Nano-Biocomposites for Food Packaging," In: Avérous L., Pollet E. (eds) *Environmental Silicate Nano-Biocomposites. Green Energy and Technology*. Springer, London. 10.1007/978-1-4471-4108-2_15.

[32] H. C. Voon, R. Bhat, A. M. Easa et al. "Effect of Addition of Halloysite Nanoclay and SiO2 Nanoparticles on Barrier and Mechanical Properties of Bovine Gelatin Films," *Food Bioprocess Technol*, vol. 5, pp. 1766–1774, 2012. 10.1007/s11947-010-0461-y.

[33] N. Bumbudsanpharoke, J. Choi, and S. Ko, "Applications of nanomaterials in food packaging," *Journal of Nanoscience and Nanotechnology*, vol. 15, no. 9, pp. 6357–6372, 2015.

[34] M. Makaremi, P. Pasbakhsh, G. Cavallaro et al., "Effect of morphology and size of halloysite nanotubes on functional pectin bionanocomposites for food packaging applications," *ACS Applied Materials & Interfaces*, vol. 9, no. 20, pp. 17476–17488, 2017.

[35] S. Ganguly, K. Dana, T. K. Mukhopadhyay, T. K. Parya, and S. Ghatak, "Organophilicnano clay: a comprehensive review," *Transactions of the Indian Ceramic Society*, vol. 70, no. 4, pp. 189–206, 2011.

[36] M. Farhoodi, "Nanocomposite materials for food packaging applications: characterization and safety evaluation," *Food Engineering Reviews*, vol. 8, no. 1, pp. 35–51, 2016.

[37] F. Sadegh-Hassani, and A. MohammadiNafchi, "Preparation and characterization of bionanocomposite films based on potato starch/halloysite nanoclay," *International Journal of Biological Macromolecules*, vol. 67, pp. 458–462, 2014.

[38] B. Biduski, J. A. Evangelho, F. T. Silva et al., "Physicochemical properties of nanocomposite films made from sorghum-oxidized starch and nanoclay," *Starch-Stärke*, vol. 69, no. 11-12, article 1700079, 2017.

[39] W. Gao, H. Dong, H. Hou, and H. Zhang, "Effects of clays with various hydrophilicities on properties of starch-clay nanocomposites by film blowing," *Carbohydrate Polymers*, vol. 88, no. 1, pp. 321–328, 2012.

[40] T. J. Gutierrez, A. G. Ponce, and V. A. Alvarez, "Nano-clays from natural and modified montmorillonite with and without added blueberry extract for active and intelligent food nanopackaging materials," *Materials Chemistry and Physics*, vol. 194, pp. 283–292, 2017.

[41] K. Marsh, and B. Bugusu, "Food packaging - roles, materials, and environmental issues," *Journal of Food Science*, vol. 72, no. 3, pp. R39–R55, 2007.

[42] C. Han, A. Zhao, E. Varughese, and E. Sahle-Demessie, "Evaluating weathering of food packaging polyethylene-nano-clay composites: release of nanoparticles and their impacts," *Nano*, vol. 9, pp. 61–71, 2018.

[43] E. Castro-Aguirre, R. Auras, S. Selke, M. Rubino, and T. Marsh, "Impact of nanoclays on the biodegradation of poly (lactic acid) nano composites," *Polymers*, vol. 10, no. 2, p. 202, 2018.

[44] H. Qin, Z. Zhang, M. Feng, F. Gong, S. Zhang, and M. Yang, "The influence of interlayer cations on the photo-oxidative degradation of polyethylene/montmorillonite composites," *Journal of Polymer Science Part B: Polymer Physics*, vol. 42, no. 16, pp. 3006–3012, 2004.

[45] T. O. Kumanayaka, R. Parthasarathy, and M. Jollands, "Accelerating effect of montmorillonite on oxidative degradation of polyethylene nanocomposites," *Polymer Degradation and Stability*, vol. 95, no. 4, pp. 672–676, 2010.

[46] S. M. Sapuan, Y. Nukman, N. A. A. Osman, and R. A. Ilyas, *Composites in Biomedical Applications* (1st ed.). CRC Press, 2020. 10.1201/9780429327766.

[47] A. Mishra, and S. Mishra, 2011. "Cellulose Based Green Bioplastics for Biomedical Engineering," In *Handbook of Bioplastics and Biocomposites Engineering Applications*, S. Pilla (Ed.). 10.1002/9781118203699.ch12.

[48] G. L. Robertson, *Food Packaging: Principles and Practice*, Third Edition (3rd ed.). CRC Press, 2012. 10.1201/b21347.

[49] C. Irwin, 2010. "Blow molding," In: K. L. Yam (Ed.), *The Wiley Encyclopedia of Packaging Technology*, third ed. John Wiley and Sons, New York, pp. 137–154.

[50] L. Pascucci, 2010. "Injection molding for packaging applications," In: K. L. Yam (Ed.), *The Wiley Encyclopaedia of Packaging Technology*, third ed. John Wiley and Sons, New York, pp. 586–594.

[51] V. Chougule, and M. Piercy, 2010. "Thermoforming," In: K. L. Yam (Ed.), *The Wiley Encyclopaedia of Packaging Technology*, third ed. John Wiley and Sons, New York, pp. 1228–1236.

[52] K. W. Shu, and M. H. Tusim, 2010. "Foam plastics," In: K. L. Yam (Ed.), *The Wiley Encyclopaedia of Packaging Technology*, third ed. John Wiley and Sons, New York, pp. 518–527.

[53] J. Dooley, and H. Tung, 2001. "Coextrusion.2," In: *Encyclopaedia of Polymer Science and Technology (1999-2014)*, fourth ed., vol. 15. John Wiley and Sons, New York].

[54] S. Farris, I. U. Unalan, L. Introzzi, J. M. Fuentes-Alventosa, and C. A. Cozzolino, "Pullulan-based films and coatings for food packaging: Present applications, emerging opportunities, and future challenges," *J. Appl. Polym. Sci.*, vol. 131, pp. 40539, 2014. doi: 10.1002/app.40539.

[55] S. Farris, and L. Piergiovanni, 2012. "Emerging coating technologies for food and beverage packaging materials," In: Yam, K. L. (Ed.), *Emerging Food Packaging Technologies Principles and Practice*. Woodhead Publishing Limited, Cambridge.

[56] M. G. Alsdorf, 2010. "Extrusion coating," In: K. L. Yam (Ed.), *The Wiley Encyclopaedia of Packaging Technology*, third ed. John Wiley and Sons, New York, pp. 742–750.

[57] R. Bakish, 2010. "Vacuum metallizing," In: K. L. Yam (Ed.), *The Wiley Encyclopaedia of Packaging Technology*, third ed. John Wiley and Sons, New York, pp. 440–444.

5 Thermal Performance Studies of a V-Type Baffle Absorber Plate Used in a Solar Air Heater

Partha Pratim Dutta and Harjyoti Das
Department of Mechanical Engineering, Tezpur University,
Tezpur, Assam, India

CONTENTS

5.1   INTRODUCTION

Solar air heating refers to the alteration of solar radiation towards useful thermal energy. Air is unable to directly absorb solar radiation effectively due to its low heat capacity. Therefore, an added process is necessary to make this possible for the transfer of energy and to deliver the output of heated air into useful means for living space or working space. The heater plate's thermal energy is transferred to a living or working space by air, which absorbs it. The methods that have been designed in order to facilitate this process are known as solar air heaters. Energy conservation comes first, followed by the investigation of active solar thermal technology and the application of a passive collection of solar energy, distribution, and storage techniques. Again, active solar technology deployment has been made easier because air-heating solar collectors may easily be added to farm building roofs. Rooftop solar collectors are constructed over an absorber made of black plastic, metal, or metal having glass coatings, for the purpose of heating air, there is a clear plastic film cover. Metal solar air heaters are constructed on farms, and range from double-glazed moderate-temperature crop drying techniques to unglazed low-temperature livestock systems.

Air temperature cannot be elevated to a higher value directly by solar radiation due to its low thermal conductivity. A metal plate captivates the solar radiation well. The hot metal plate's heat is transferred to the air that is moving across it. This device is referred to as a solar air heater. Solar air heaters have several applications like textile, food, paper, and household applications. So as to reduce the carbon footprint triggered by using conventional heat sources like fossil fuel, and industrial processes, solar air heaters may be utilized to some extent to produce an alternative and sustainable means of heat energy. Using solar air heaters (SAH) to heat spaces in commercial plants plus in residential areas thus reducing the use of other potential pollution approaches of heating. It may be used in the drying process of crops (tea, black pepper, green tea, corn, etc.), laundry drying, and many other drying applications.

Because of the lower value of air's heat capacity, conventional flat solar air heaters have poor performance. So, the absorber plate can be artificially roughened, which is thought to be a practical and cost-effective way to improve thermohydraulic performance. The plate has been subjected to tests for various sorts of roughness to find the finest roughness geometry to improve its performance. The primary emphasis of numerous researchers is the optimum suited roughness geometry for the absorber plate that could provide the best performance in relation to the mass flow rate employed by solar air heaters. SAHs accumulate solar energy and transform it into usable form by increasing the air temperature passing through them. Then, this is utilised for food preparation, drying, and heating of space, amongst other things. Since no air can freeze, leakage of air is less severe than with other working agents, and SAHs are less corrosive, they have numerous advantages over conventional heating systems. These advantages set them distinct from other similar systems. However, due to the minimal efficacy of such devices, researchers chose to learn more about SAHs. The efficiency of solar devices may be considerably increased by making alterations to the design and construction, the use of vortex generators and other tools to promote heat transmission, the roughening of the absorber plate, the use of PCMs and nanofluids, and other techniques. A SAH comprises three major components: a plate (absorber), an insulated enclosure, and an air blower. The collector's surface is where solar energy is reflected and absorbed. The air travelling beneath the collector is heated because of the collector's heat. A rectangular collecting chamber is securely positioned inside the absorber plate. To reduce top losses, the collector box's top portion is covered with glass. A blower is used to drive air via an absorber plate. Thermocouples are linked to the absorber plate so that the temperature can be measured. The insolation is also measured with a pyranometer. These are simple systems that are employed in low-temperature applications. Typically, there is no way to monitor the sun's orientation with respect to the location under consideration. SAH, absorber plates are the most important component. Hence these are coated in a dull black colour to produce an almost black body that will absorb as much radiation as possible. Because of the sub-layer (laminar) formation above the heater plate, a simple SAH performs poorly (absorber). This provides confrontation to moving air while also creating a barrier, preventing air from touching the surface that is heated. A large temperature differential is required for a good amount of heat transmission, which is not attainable in normal air heating by solar energy. As a result, the researchers used a variety of approaches to optimize heat transmission so that the system's performance might be improved and implemented for better performance. A few techniques are applied to boost the system's performance like putting artificial roughness on the absorber surface. Based on the temperature obtained and the efficiency reached, solar air heater collectors are divided into different categories. Both diffuse and beam solar light can be collected by flat plate collectors, which then transform it into low-temperature heat energy.

In a conventional SAH, air can pass through by laying one absorber plate on top of another plate. Over the plate (absorber), one or additional transparent covers are used, and insulation is placed in a container. It is given at the bottom and sides to limit losses. In three different ways, the absorber plate's corrugation improves the heat transfer process: (1) the region for heat transfer is expanded; (2) subjecting to the flow direction turbulence may rise; and (3) it is possible to directionally select the surface if oriented appropriately. Both a recycled flow and no recycled flow may be used with the double-pass airflow arrangement. In terms of precise airflow estimation, the performance of the collectors has significantly improved during counter-flow activities. However, bending-related losses, underdeveloped flow, and other factors are unacceptable. A modified form of the conventional SAH is the SAH with fins. This SAH has fins lining the airflow path. A normal flat plate with the same size has a lower heat transfer region than the finned absorber surface. SAH with baffles is a useful technique to increase heat transfer when compared to a standard flat plate SAH. The airflow path in this instance has baffles. Heat transfer is accelerated by the baffles with the creation of turbulence on the absorber plate. The porous SAH includes, among other things, wire mesh, limestone, and gravel. Comparable heat is conducted throughout the day by this porous bed-type SAH when solar radiation is weak or nonexistent while absorbing heat energy during high-intensity solar radiation. The performance of the SAH is significantly affected by the absorber plate's size and design. An increase in the rate of heat exchange between the solar collector and the air flowing over it improves the SAH's thermal efficiency. The disadvantage is that when pressure reductions grow, the fan or blower must work harder to move air through the air heater. This suggests that some changes must be made to the fundamental mechanism in order to speed up the rate of heat transmission. The system's performance is improved owing to the output air temperature increment with mass flow rate or with various absorber geometries. Fins are one of the modifications made to an absorber. Solar heat collectors come in two different varieties: glazed and unglazed. Unglazed collectors lack the glass covering that is commonly referred to as glazing. To increase the lifespan of the system, they are often manufactured of durable rubber or plastic that has been treated with an ultraviolet (UV) light inhibitor. Unglazed collectors are less costly than glazed collectors due to cheaper materials and a simple design. In cooler regions, indoor pools also employ unglazed collectors because they enable water to return to the pool when it is not being used. Installing unglazed collectors is cheaper than installing more costly glazed collectors, even if the user must temporarily switch the system off during colder weather. Many individuals believe those costly glass coverings are required and that unglazed covers are useless. Unglazed coverings are not only adequate for pools in the open air, but they are also less expensive. The heat from the pool is usually lost by evaporation, convection, and direct conduction from the water's surface. A large amount of heat is lost as a result if the air temperature nearby is significantly lower than just the temperature of the water. As a result, maintaining a comfortable air temperature may help to support the water temperature. A glazed collector has a tempered iron glass covering and is manufactured from copper tubing on an aluminium plate, and they are more costly. Glazed collectors with heat exchangers transport fluids more effectively than unglazed collectors during colder seasons. This allows them to capture sunlight much more effectively. They are suitable for usage throughout the year and in various conditions. If utilized in cooler temperatures, unglazed and glazed collector systems must include freeze protection. Unglazed air collectors are solar air heating systems that consist of a metal absorber plate without any glass or glazing on the top. The primary use of unglazed solar collectors is to heat ambient air. An absorbing medium is used in the glazed air collector. Polymer glass may be used for capturing solar radiation. A simple solar air collector is made out of a five-sided insulated container, a selected surface, and an absorber material to trap and preserve solar energy. Unglazed air collectors are solar air heating systems that consist of a metal absorber plate without any glass or glazing on the top. The primary use of unglazed solar collectors is to heat the ambient air and it is not meant for heating indoor building air.

Through convective heat transmission, the absorber plate transmits thermal energy to the air. Thereafter, for goals of space heating as well as process heating, the heated air is transferred. It is also transferred to the building space as required. While researching solar air heaters, it is important to be aware of both their high advantages and disadvantages. The few advantages are; since air is used directly in the process, heat transfer from one fluid to another is not necessary. It is straightforward and convenient to utilize the system. A significant problem with solar water heaters is corrosion. As there is no issue with air leaking from the duct, solar air heaters are not affected by this issue. There is no working fluid freezing. The pressure inside the collector never rises above a certain point. As a result, air heaters are more user-friendly and may be constructed using less expensive materials than solar water heaters. The first and most serious problem is that air has poor heat transmission capabilities. Improved heat transmission needs more caution. Another disadvantage is the low air density, which necessitates handling enormous volumes of air. The low thermal capacity of air prevents its use as an effective fluid storage medium. Solar air heaters may be costly if they are not appropriately designed.

Artificial roughness works well for increasing the solar air heaters' capacity for transferring heat. Researchers have always aimed to increase the performance of V-rib geometry. Due to the current energy scarcity, there is a demand for equipment such as compact solar air heaters, which have low power consumption, require little maintenance, and serve multiple purposes. A study was conducted for the coordinates 26°6528' N, 92°7926' E. (Tezpur University). This study's objective is to build and simulate a baffled solar air heater for the climate of Tezpur, Assam. A comparison was made between a continuous V-type baffles absorber plate and a discrete V-type baffles absorber plate by estimating thermo-hydraulic efficiencies.

5.2 LITERATURE REVIEW

The important unit is the solar air channel that transforms solar radiation into useful thermal energy. Considering the rising demand along with the decreased input of primary energy from non-renewable sources, a solar heater has been considered an alternative. These were created through a straight transformation of solar energy towards heat produced by a means of transportation, i.e., internal energy. The important device with the help of which solar energy is converted to a useful temperature and heat energy is the solar air channel. Solar air collectors can be utilized for different uses like industrial drying of food, pharmaceutical, paper, pulp, area heating at home or any place, and timber seasoning from time to time. These can also be used for drying or curing clays or concrete for buildings. Baffles of certain heights that function as turbulators are used to make turbulence throughout the flow, thereby increasing the heat exchange rates. Because the fluid and heated absorber are better mixed, the heat transfer rates are increased. However, substantial pressure drops are associated with baffles [1]. Rajaseenivasan et al. looked at circular and V-shaped ribs over Nusselt number (Nu_{rs}) and friction factor (f_{rs}) with Reynolds number (Re) ranging from 6000 to 12,000 over a channel for a SAH. There's an enhancement in Nu_{rs} to be augmented when the Re increased in the heated absorber plate, and extreme data of Re = 11,615 was reached [2]. Bovand et al. investigated and numerically found the Nu_{rs} and f_{rs} in a solar air channel and noted the effect of using porous materials [3]. Gawande et al. studied using simulations and mathematical modelling techniques to evaluate thermo-hydraulic performance and effective efficiencies at 20° rib angled for a roughened solar air conduit. There was a finding that the highest Nu_{rs} and f_{rs} were obtained at $P_b/H = 7.143$, $H_b/H = 0.042$, and Re = 15,000 [4]. Nuntadusit et al. experimentally found that performance improvement in a wind channel might be achieved with the baffles cut for the improvement of Nu_{rs} [5]. Hu et al. investigated a straightforward design with interior baffles solar air collectors. The presence of the baffles was found to speed up heat transfer. Again, reduced radiation loss led to greater effectiveness [6]. Ravi and Saini experimented with various discrete broken multi v patterns and ribs that were staggered to create unevenness on two sides of the absorber. They found a relationship between friction factor in forced convective type air heaters as well as the Nusselt number for counter flow [7]. Kumar et al. did an experimental analysis of the impact of different variables, friction factor and Nusselt number correlations with V-type baffles [8]. Priyam et al. observed solar air heater's efficiency with finned absorbers analytically. Two wavy fins were placed transversely and fastened to the solar collector of the channel. On the top side of the absorber, the fluid was subjected to a constant heat flow, and the bottom surface was thermally insulated [9]. Skullong, et al. studied and developed the collector efficiency factor for solar collectors. They investigated the improvement of heat transport experimentally over a channel of an air heater having an arrangement of wavy grooves and winglets [10]. Experimental and modelling studies have reported different improved geometry solar air heater's absorber plates [11–19]. Dutta et al. [20–22] studied corrugated heat exchangers and solar air heaters for use in solar drying. Dutta et al. experimented with *Garcinia pedunculata* drying by means of a corrugated solar dryer through free convection [23]. Goel et al. thoroughly examined the history, basics, and most recent improvements in solar thermal air heating technology. Various solar collector designs are described and reviewed, including evacuated tubes, flat plates, numerous passages, cross-sections of the flow passages, and so on. The use of vortex generators, fins, baffles, and other devices to enhance SAH performance is described. To determine the thermos-hydraulic performance parameters and choose the best configuration for applications, a comparative performance study has been done [24]. Arunkumar et al. examined the effectiveness of solar air heaters with added spring-like fins just below the absorber surface. Across the entire range of Re, it is observable that the thermo-hydraulic and Nu's are high enough for a 0.133 helicoidal pitch ratio. As the value of Nu increases, the ratio of springy wire diameter increases. On the other hand, the thermo-hydraulic rise factor grows until it falls because of increased flow resistance as soon as the springy wire size ratio reaches 0.093. As the helicoidal spring diameter proportion rises, the Nusselt number decreases. The intensity for thermo-hydraulic is significantly higher for lower Reynolds numbers, with 0.06 helicoidal spring diameter ratios at 1.268 [25]. Kalash et al. observed the effectiveness of a double-pass SAH utilizing simulated and experimental dynamic fluid simulation. Ten aluminium pipes were used to build 3 double-pass SAHs for the testing. The sides and bottom base were insulated with glass wool. The results showed that as time passed, the temperature and solar flux increased in the morning and subsequently declined at noon. The stored thermal energy constantly increased during the day and diminished at night. The air velocity was also discovered to cause vortices in the collection [26]. Vora et al. thoroughly investigated the development of passive enhancement techniques used for solar air heaters. Inserts featuring flanges, fins, perforated surfaces, baffles, and wire mesh were employed to enhance the efficacy. Both laminar and turbulent flows show the inserts. The thermo-hydraulic behaviour of an insert is significantly influenced by the flow conditions and insert configurations [27]. Khanlari et al. studied the greenhouse dryer's (GD) thermal performance. It was enhanced by adding a reasonably priced and simple-to-use solar air heater made of tube (T-SAH). Here, a straight frontward tube-type SAH (T-SAH) has been developed, investigated, and built. Again, the primary goal to invent this sort of solar air heater was to create a more ecologically friendly and alternative air heating technology. Apricot samples were dried in the built-in dryer. The rates of airflow in ranges were 0.010, 0.013, and 0.015 kg/s. The results of the experiments showed that combining T-SAH and GD significantly shortened drying time. The average SAH's efficiency was also found to be in the range of 45.6%–56.8% [28]. Singh et al. used phase change material to examine four possible arrangements of the two distinct absorber plates for a solar air heater with (PCM). The weather in Moradabad, at longitude 78°47' E and latitude 28°50' N was utilized to assess improved SAH models using PCM. The difference in temperature between the exhaust air and outside air of SAH using PCM-filled small cylindrical tubing is typically between 2°C and 9°C when compared to SAH with a flat plate solar absorber using forced convection. Among all the examined setups, configuration 4 was shown to be the most effective in terms of outlet temperature. The temperature was around 48°C, with around 9.8 hours per day of solar radiation. The daily efficiency of 66% was ideal for use in agricultural drying and space heating [29]. Parsa et al. explored turbulent convective heat transmission with special staggered cuboid baffles put on the absorber of the heater duct absorber. Using a technique for optimizing in accordance with the Taguchi experimental design methodology, the properties of heat transport and pressure loss were assessed, and the ideal geometry and arrangement of the baffle were established. The numerical findings show that the recommended optimum design outperforms other designs. Efficiency and thermo-hydraulic efficacy are mostly influenced by comparative baffle height (c/H) and comparative baffle pitch (s/H). At Reynolds numbers 5080, 7620, and 10,160, correspondingly, the optimal baffled SAH does have a thermo-hydraulic performance factor having values of 3.43, 2.80, and 2.38. The suggested baffling's thermo-hydraulic performance factor improves the performance of the best design in the literature by up to 17.5%, based just on Reynolds number [30]. Niyonteze et al. studied significant efficiency, achieved when all six proposed metaheuristic approaches were used to choose the ideal design set and operational elements for solar air heaters. Depending on prior studies along with the findings of this work, for the same investigated problem, ETLBO is certainly more competitive when it comes to SAH optimizing than ABC, GA, PSO, SA, and TLBO. Again, grounded on the 6 cutting-edge metaheuristic approaches discussed, some conclusions and suggestions for the purpose of enhancing SAHs are offered. The first attempt is to introduce recent advancements in metaheuristic applications to a broad audience in this study. In order to conduct additional research and enhance the suggested and further suggested methods producing the greatest act aimed at a variety of SAHs, researchers might make use of SAH optimization approaches [31]. Khatri et al. tried out SAHs with cylindrical porous fins made of aluminium wire mesh and an arching wavy absorber panel. The absorber plate had been created and developed by them. Air turbulence and vortex formation increased as a result of the arching absorber plate design. A comparison of three different fluid flow velocities was done for various layouts. With a 5 m/s fluid flow rate, the wavy absorber plate with fins that are arched provided superior thermal efficiency over other variations. The comparable change in temperature was roughly (55–70)°C [32]. Singh studied numerically and experimentally the thermal efficiency of a packed bed SAH made of porous serpentine wavy wire mesh. The best thermohydraulic and thermal efficiencies were found to be 74% and 80%, respectively. In comparison to a single pass, efficiencies of SAHs with the porous double passage, and serpentine beds were around 18% and 17% higher, respectively. The numerical results demonstrate that serpentine-packed bed SAHs have a maximum thermohydraulic performance improvement of 24.33% in comparison with flat-packed bed SAHs [33]. Shetty et al. conducted a study for absorber plate configurations with numbers of vents of 24, 36, and 54 vent widths of 5, 8, 10, and 12 mm, accordingly. The results of the CFD analysis were compared to those of the experiments. When contrasted with the basic model not having the absorber surface, the configuration that has 8 mm diameter vents and 36 vents generated an average gain in thermal efficiency of 23.33%. The rise in vent diameter affects the collector's thermohydraulic efficiency. According to the study, the rounded geometry and vented type absorber generate vortex formation of a vortex, which increases turbulence-induced heat transfer. [34]. Antony et al. present a numerical analysis so as to calculate performance metrics like the thermohydraulic performance parameter plus thermal enhancement factor for Reynolds numbers between 3000 and 24,000, for a flat plate SAH using stepwise cylindrical turbulators, positioned underneath the solar collector plate. The results showed that the flow became very turbulent due to the cylindrical stepped turbulators. This flow behaviour with the flow separation around the staggered generators improves performance. With more steps, the thermal enhancement factor (TEF) has greatly risen. A Reynolds number of 15,000 results in a peak TEF of 1.14. The study generates a maximum thermos-hydraulic performance parameter of 1.49 or a Reynolds number of 18,000 [35].

5.3 MATERIALS AND METHODS

For investigating the effects and behaviour of broken V-type baffles and continuous V -type baffles over the Nu_{rs} and f_{rs} in a duct with air as a medium, ASHRAE standard 93–77 recommended design criteria were used to create an experimental configuration similar to that previously used by Kumar [8]. The channel consists of three sections, 500 mm in length; the inlet section test section is with length 1200 mm. The length of the exit sector is 300 mm. The total channel of air is 2000 mm in length and has a width of 300 mm over the cross-section, and the channel height is 30 mm. Channel insulation of 50 mm thick is provided with thermal conductivity of 0.037 W/mK. The orifice plate was pre-calibrated in the system to measure the flow rate of air. An air blower was employed to force the air through the duct. The orifice meter had a coefficient of discharge of 0.62 for calculating the mass flow rates. Temperatures were recorded using PT-100 type thermocouples at various locations on the absorber plate. The pressure was found through a micro-manometer connected along the solar air heater's span. The average pressure drop data was used to estimate the average friction factor. Periodically, various properties were analysed, including the temperature of the airflow at the intake and outlet, the temperatures over the plate, and solar radiation. These measured parameters are essential in investigating Nusselt number, friction factor, and thermo-hydraulic performance metrics. For the experimental analysis, the setup was developed and analysed by the use of the following components: Table 5.1 represents a list of components and instruments used for experimentations, and Table 5.2 represents the operating and geometrical parameters of the baffled solar air heater.

Figure 5.1 represents a different view with geometrical dimensions of the discrete V-type baffle absorber plate. Figure 5.2 represents different views of continuous V-type baffle SAH with absorber plates. Figure 5.3 is a 3D model of the complete experimental setup. Figure 5.4 is a 3D view of the broken baffle solar air heater.

5.4 GOVERNING EQUATIONS

The following assumptions do the simplification of the governing equations:

   i. Incompressible flow of air (1D).
  ii. Constant thermal and physical properties.
 iii. Steady flow.

TABLE 5.1
List of Components Used for the Experiment

Sl. No.	Items	Dimensions
1.	Enclosure wooden box	2010 mm × 310 mm × 120 mm
2.	Glass cover	2000 mm × 300 mm × 4 mm
3.	Absorber plate	2000 mm × 300 mm × 2 mm
4.	Air blower	6 $speed$, 300 W
5.	Thermocouples	− 10°C to 120°C
6.	Pyranometer	0–1999 W/m^2
7.	Orifice plate	0.10–40.0 m/s
8.	Micromanometer	0.1–100 hPa

TABLE 5.2
Operating Parameters and Geometry Parameters Used for the System

Sl. No.	Parameters	Values
1.	Collector width (W), m	0.3
2.	Glazing covers used (N)	1.0
3.	Collector length (L), m	2.0
4.	Transmittance-absorptance value($\tau\alpha$)	0.80
5.	Glass emittance value(ε_g)	0.880
6.	Plate emittance value (ε_p)	0.90
7.	Depth of glass cover (t_g), m	0.0040
8.	Thermal conductivity (Insolation) (K_i), W/mK	0.0370
9.	Thickness of insulation (t_e), m	0.050
10.	Relative roughness pitch (P_b/H_b)	3
11.	Relative roughness height (H_b/D_{hd})	0.275
12.	Baffle height (H_b), m	0.015
13.	Rise in temperature ΔT, K	10–20
14.	Environment temperature, K	300
15.	Velocity of wind (V_w), m/s	5
16.	Insolation (I), W/m^2	700–900

The governing equations used are given below:

The pressure drop is measured using the micro-manometer. For estimating the air stream flow rate Equation (5.1) is used:

$$m_a = C_{do}A_o \left[\frac{2\rho_a (\Delta p)_0}{1 - \beta^4} \right]^{0.5} \tag{5.1}$$

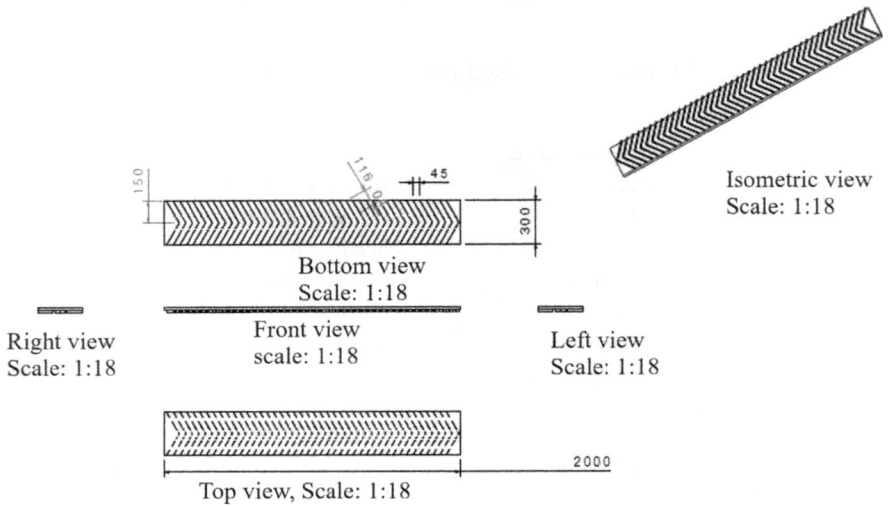

FIGURE 5.1 Discrete V-type baffle absorber plate views (All dimensions are in mm).

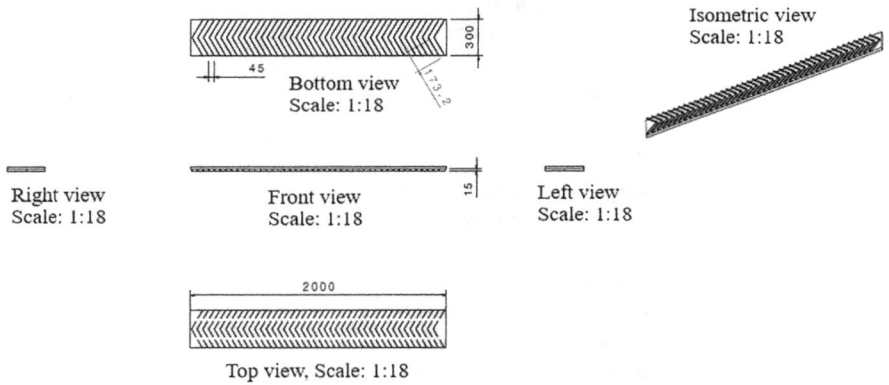

FIGURE 5.2 Continuous V-type baffle absorber plate draft (All dimensions are in mm).

The air velocity (V) is found using the value of m_a, and it is given by Equation (5.2).

$$V = \frac{m_a}{\rho_a WH} \qquad (5.2)$$

Determination of equivalent hydraulic diameter (D_{hd}) is given by Equation (5.3).

$$D_{hd} = \frac{4WH}{2(W + H)} \qquad (5.3)$$

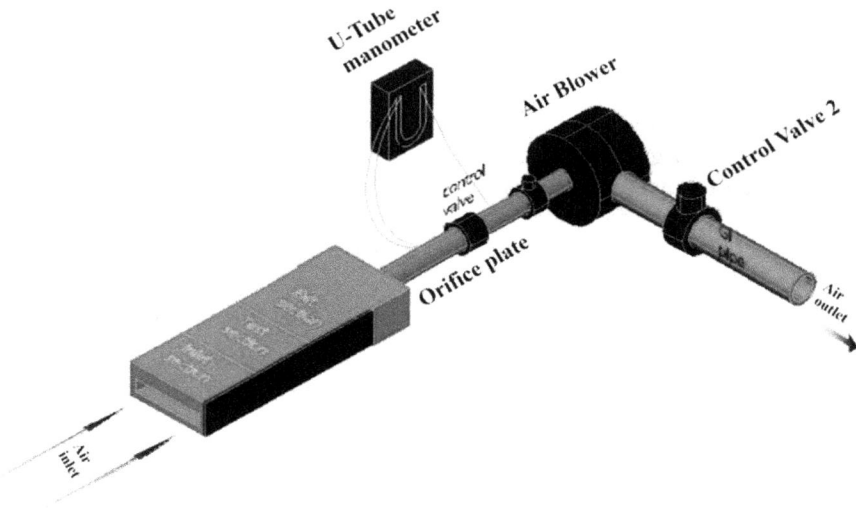

FIGURE 5.3 3D model of the complete experimental setup.

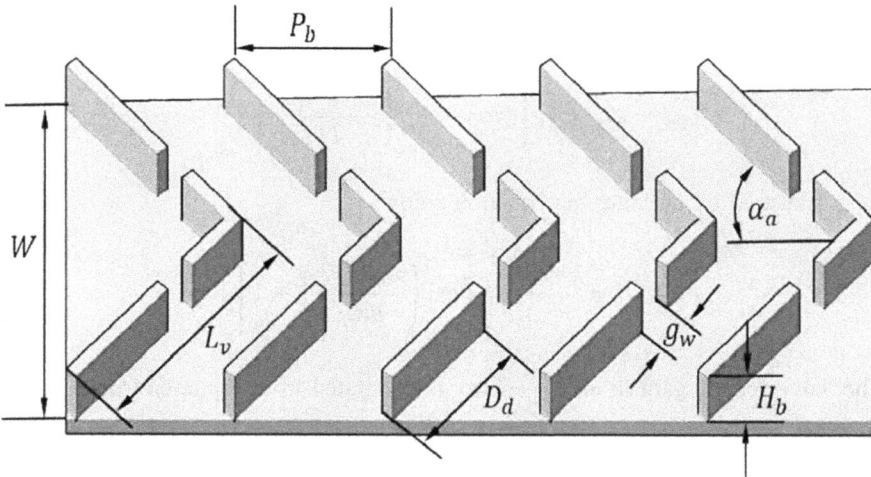

FIGURE 5.4 Broken V-type baffle.

Reynolds number (Re) of the air flow is found out from Equation (5.4).

$$R_e = \frac{VD_{hd}}{\upsilon} \tag{5.4}$$

Darcy equation is used for friction factor Equation (5.5).

$$f_{rs} = \frac{2(\Delta p)_d D_{hd}}{4\rho_a L_t V^2} \tag{5.5}$$

The mathematical procedure for the following was presented by Chamoli et al. [19]. The useful energy gain was found in relation to the mean absorber temperature by means of the following equations pertaining to the air flowing through the SAH.

$$Qu = A_p [I(\tau\alpha) - U_L(T_p - T_a)] \tag{5.6}$$

Thermal efficiency and collector heat removal factor for the SAH can also be written as,

$$n_{th} = \frac{Qu}{I.A_p} = F_R \left[(\tau\alpha) - U_L \left(\frac{T_i - T_a}{I} \right) \right] \tag{5.7}$$

$$F = \frac{\dot{m}C_p}{U_L} = \left[1 - \exp\left(\frac{U_l A_p F_p}{\dot{m}C_p} \right) \right] \tag{5.8}$$

Thermal efficiency can also be taken by means of the exit temperature of the fluid as,

$$n_{th} = F_0 \left[(\tau\alpha) - U_L \left\{ \frac{T_0 - T_i}{I} \right\} \right] \tag{5.9}$$

F_0 is analyzed by utilizing the equation below:

$$F_0 = \frac{\dot{m}C_p}{U_L.A_p} = \left[\exp\left(\frac{U_L A_p F'}{\dot{m}C_p} \right) - 1 \right] \tag{5.10}$$

The value for the gain in usable energy is computed from Equation (5.6)

$$Q_u = \dot{m}C_p (T_0 - T_i) \tag{5.11}$$

Thermal efficiency is calculated from Equation (5.12).

$$n_{th} = \frac{\dot{m}C_p (T_0 - T_i)}{I.A_p} \tag{5.12}$$

It was seen that the strong functions are the average friction factor and average Nusselt number for roughness geometry parameters. From the experimental data that existed previously, a regression analysis was done to find a statistical Nusselt number correlation. The correlation for Nu for the discrete V-type SAH could be written as [8]:

$$Nu_{rs} = 0.261 \ Re^{0.679} \qquad (5.13)$$

Similarly, the final correlation for Nu for the continuous V-type solar air heater is written as:

$$Nu_{rs} = 0.049 \ Re^{0.835} \qquad (5.14)$$

Similarly, based on the regression analysis adopted, the statistical friction factor correlation of the experimental data was found. The statistical friction factor correlation found for the discrete V-type solar air heater resulting based on regression analysis can be written as,

$$f_{rs} = 0.158 \ Re^{-0.10} \qquad (5.15)$$

Similarly, the final friction factor correlation for the continuous V-type can be written in the form of,

$$f_{rs} = 0.113 \ Re^{-0.07} \qquad (5.16)$$

Comparison between the value of Nu_{ss} and f_{ss} were already calculated through experimental outcomes and was made with the values found for Nu_{ss} from the Dittus-Boelter equation and f_{ss} from the modified Blasius equation [5.8].
The Dittus-Boelter equation gives the Nu_{ss} for the smooth channel as:

$$Nu_{ss} = 0.02 \ Re^{0.8} Pr^{0.4} \qquad (5.17)$$

The modified Blasius equation gives the f_{ss} for a smooth channel as:

$$f_{ss} = 0.085 \ Re^{-0.25} \qquad (5.18)$$

Thermal-Hydraulic Performance Parameter:

Nu_{rs} and f_{rs} characteristics studies show that an improvement in Nu_{rs} always leads to friction power loss because of an increase in the $f_{rs,}$ on the other hand. Hence, a decision for the geometry parameters is important for an effective increase in Nu_{rs} with the lowest f_{rs} cost. Lewis [17] proposed a relation for the thermo-hydraulic parameter. It is also known as the efficiency parameter 'η_p'. For the very reason of achieving the higher Nusselt number considering the same pumping power requirement for a roughened SAH in comparison to that of smooth and is defined as:

$$\eta_p = (Nu_{rs}/Nu_{ss})/(f_{rs}/f_{ss})^{0.33} \qquad (5.19)$$

The values of U_L and F_0 are required for estimating the thermal efficiency from Equation (5.9). For this purpose, the following equations are used. The overall heat loss coefficient is determined by adding the top, bottom, and edge loss coefficients, as shown in the equation below.

$$U_L = U_t + U_b + U_e \tag{5.20}$$

The equation following can be used to compute the bottom loss coefficient,

$$U_b = \frac{k_i}{t_i} \tag{5.21}$$

The equation following can be used to compute the edges loss coefficient,

$$U_e = \frac{(W + L) \times L_1 \times k_i}{W \times L \times t_e} \tag{5.22}$$

The proposed equation by Kline [18] for the empirical equation of the top loss coefficient is given below,

$$U_l = \left[\frac{N}{\left(\frac{C}{T_m}\right)\left(\frac{T_m - T_a}{N+f}\right)^{0.252}} + \frac{1}{h_w} \right]^{-1} + \left[\frac{\sigma(T_m^2 + T_a^2)(T_m + T_a)}{\frac{1}{\varepsilon_p + 0.0425N(1 - \varepsilon_p)} + \frac{2N+f-1}{\varepsilon_g} - N} \right] \tag{5.23}$$

Where, N= number of glasses covers.

$$f = \left(\frac{9}{h_w} - \frac{30}{h_w^2} \right)\left(\frac{T_a}{316.9} \right)(1 + 0.091N)$$

$$C = \frac{204.429(cos\beta)^{0.252}}{L^{0.24}}$$

$$h_w = 5.7 + 3.8 \times V_w$$

β = collector tilt from horizontal
ε_g = emittance of glass
ε_p = emittance of the absorber plate
T_a = ambient temperature (K)
T_m = mean plate temperature (K)
h_w = wind heat transfer coefficient (W/m^{2}K)
L = spacing between plates (m)

For finding the top loss coefficient, the initial plate temperature is approximated using the following equation below,

$$T_p = \frac{T_0 - T_i}{2} + 10°C \qquad (5.24)$$

where,

$$T_0 = T_i + \delta T \qquad (5.25)$$

By using Equation (5.6), the useful energy gain is found by the equation given below,

$$Q_{u1} = A_p [I(\tau\alpha) - U_L(T_p - T_a)] \qquad (5.26)$$

The useful energy gain evaluated in Equation (5.26) is validated using the equation below while computing,

$$Q_{u2} = A_p F_0 [I(\tau\alpha) - U_L(T_0 - T_i)] \qquad (5.27)$$

The following formulae are used to calculate the heat removal factor and collector efficiency factor,

$$F_p = \frac{h}{h + U_L} \qquad (5.28)$$

$$F_0 = \frac{\dot{m}C_p}{A_p U_L}\left[1 - exp\left(-\frac{A_p U_L F_p}{\dot{m}C_p}\right)\right] \qquad (5.29)$$

The equation below provides the value of the convective heat transfer coefficient,

$$h = \frac{N_u K}{D} \qquad (5.30)$$

Between Q_{u1} and Q_{u2} the absolute difference is calculated and if found 0.1% greater than that of Q_{u1}, then the plate temperature is replaced with a new value from the correlation given below,

$$T_p = T_a + \frac{I(\tau\alpha) - Q_{u2}}{U_L} \qquad (5.31)$$

The steps followed above were repeated in a loop unless the terminal criteria is found in the required range. The average of Qu1 and Qu2 is the useful heat gain Qu that is taken for calculating the thermal efficiency and defined as,

$$\eta_{th} = \frac{Q_u}{I \times A_p} \eta_{th} = \frac{Q_u}{I \times A_p}. \qquad (5.32)$$

5.5   RESULTS AND DISCUSSION

The variation of thermal enhancement factor depending on Re aimed at discrete and continuous V-type baffles SAH is shown in Figure 5.5. It shows the thermo-hydraulic performance parameter or thermo-hydraulic enrichment factor variations corresponding to Re. One could notice that the discrete broken V-type baffles at 60° angle of attack result in improved thermohydraulic efficacy than the continuous V-type baffles and the smooth plate. The discrete broken V-type baffles with 60° give good flow behaviour and efficient heating effect along with the collector plate that cannot be attained when using the smooth flat plate for heating the air. With the increase in Re values across the board, the enhancement factor as well as thermohydraulic performance inclines to decline. Maximum thermo-hydraulic performance parameter of 3.12 is provided by these discontinuous V-type baffles at comparative roughness values and consistent baffle parameters (D_d/L_v = 0:67; g_w/H_b = 1:0; H_b/H = 0:50; P_b/H = 1:5; and α = 60°) and at Reynolds number being 3,000. The thermal-hydraulic performances of the two were compared. The effects of V-type baffles on the absorber plate on friction factor and Nu stood investigated and compared with similar results obtained using a smooth channel alone. The research was performed on four various mass flow rates, with Re (3000, 6000, 9000, and 12,000).

The Nusselt numbers obtained under different Re for all different cases are presented in Figure 5.6. The variation of the Nu vs. Re for the discrete V-type baffle, continuous V-type baffles, and smooth SAH is shown and this one proves that the presence of the turbulators or baffles gives a considerable amount of heat transfer enhancements in a similar manner that can be seen in case of a smooth channel. Along with the rising value of Re, Nu also increases in value. This is because

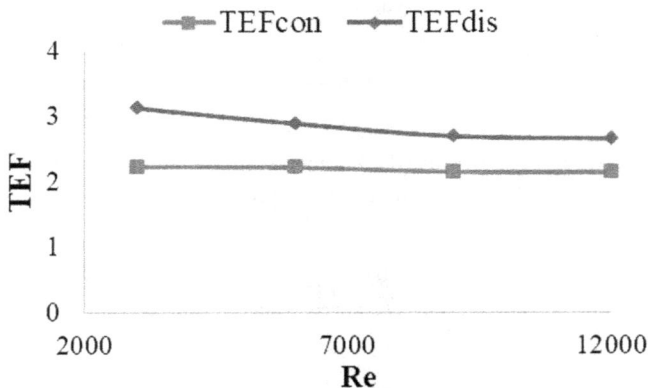

FIGURE 5.5   Thermal enhancement factor variation with Reynolds number for discrete and continuous V-type baffles solar air heater.

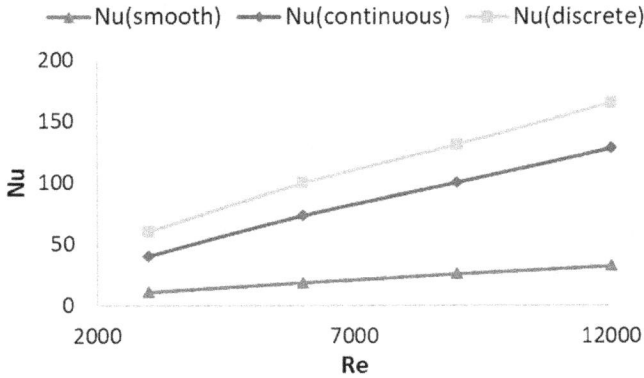

FIGURE 5.6 Variation of the Nu vs. Re for the discrete V-type baffle, continuous V-type baffles, and smooth solar air heater.

periodic V-type baffles restrict the thickening of the boundary layer development and then increase the turbulent flow of fluid. It is worth noting that the turbulence in the discrete V-type baffles is more than in the continuous V-type baffles. Interrupting the flow and changing its direction are the primary reasons for high mixing levels.

Figure 5.7 depicts the Re effect on f for broken v-baffle (D_d/L_v 0:67; g_w/H_b 1:0; H_b/H 0:50; P_b/H 1:5, and $\propto$ 60°), continuous V-type baffles and smooth wall channel. The effect of using v-baffles on a fall in isothermal pressure may be seen through the tested region of the channel. As can be observed, continuous baffles have a lower f value than broken V-type baffles. The streaming of air in the obstacles creates turbulence in the flow. Recirculation twists are established due to hindrances formed by baffles; hence, the high-pressure drop may be perceived. In this figure, it is clear that by means of the V-type baffles, this leads to a rise in friction factor when compared with smooth channels. The attributes are higher surface area, flow blockage, and reverse flow due to baffles. Again, the ratio of Nusselt numbers, Nu/Nu_{ss} (The ratio of roughened and smooth channel Nusselt numbers), varies against the Reynolds number. The Reynolds number has increased from 3000 to 12,000 while the Nusselt number ratio seems to be roughly unchanged indicating for both smooth and baffled channel Nusselt number increase at similar rates. For discrete V-type baffles, the average Nusselt number is about 5.36 times compared to the Nusselt number of flows over a smooth channel. The figure was 3.94 times the smooth channel flow for using a continuous V-type baffle. For both continuous and discrete V-type baffles, the friction factor ratio value changed with the Reynolds number. According to the experiment, with the increasing value of the Reynolds number as well as the blockage ratio, the proportion of friction factor appears to rise. While utilizing discrete V-type baffles, the average friction factor ratio is 6.97, and when using continuous V-type baffles, it is 6.12. From the result, it is clear that the use of a low blockage ratio helps decrease the pressure loss by a considerable amount.

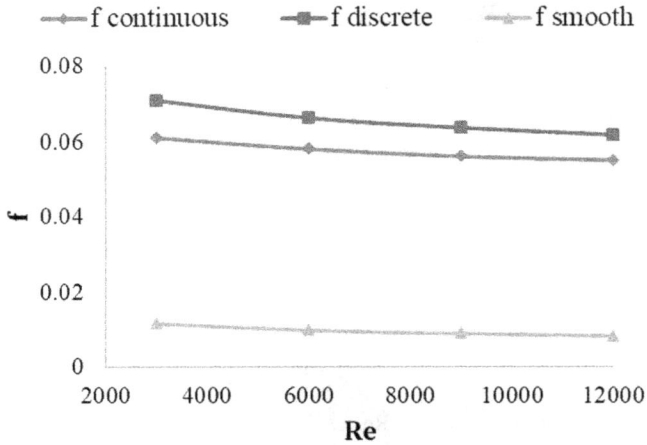

FIGURE 5.7 Variation of the f vs. Re for the discrete V-type baffles, continuous V-type baffles, and smooth solar air heater.

Thermal efficiency variations for the discrete V-type baffles and the continuous V-type SAHs with different values of *Re* at different solar insolation values are illustrated in Figure 5.8 and Figure 5.9. Efficiency rises when solar insolation values increase. The efficiency is maximum for solar radiation at 900 W/m². Thermal efficiency and useful heat intake are reduced as a result of the solar insolation's declining value, as would be expected.

As the inlet temperature of air increases, the collector temperature also increases. Therefore, with an increasing inlet air temperature, the efficiency falls. With a rising value of air velocity, the inlet and outlet air temperatures approach each other because of an increase in convective heat transfer from the plate to the fluid and a lesser residence time of air for transferring heat from the absorber plate. Figures 5.8 and Figure 5.9 show the differences in the thermal efficiency of solar air heaters

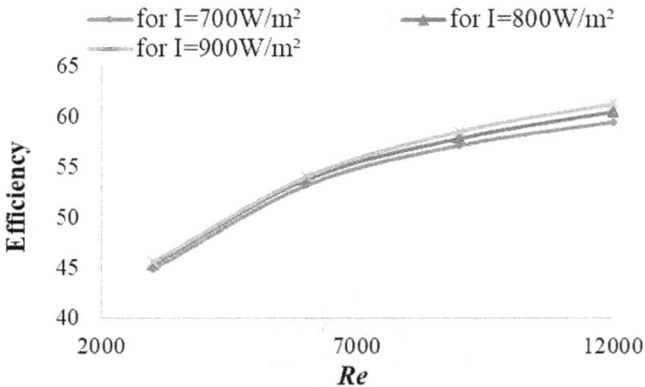

FIGURE 5.8 Efficiency variation with respect to *Re* for different values of insolation for discrete V-type baffles solar air heater.

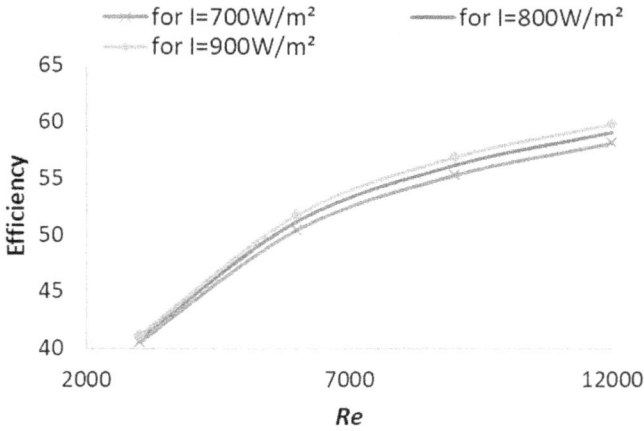

FIGURE 5.9 Efficiency variation with respect to Re for different insolation values for continuous V-type baffles SAH.

with discrete V-type baffles, continuous V-type baffles, and smooth or without baffles at various insolation values with regard to Re. Because the useful heat gain is higher for higher insolation values, the efficiency value is higher for the higher solar radiation for a particular air mass flow rate. More turbulence is created when it comes to the discrete V-type baffles, and it induces the flow separation because of the roughness of the baffles and the secondary flow generation heated up more along with the baffles. It again mixes with the primary flow, thereby breaking the laminar sub-layer formation, starting higher heat transfer. Figure 5.10 displays the efficiency variant with Re for solar air heaters with smooth or no baffles, continuous V-type baffles, for fixed insolation values of 900 W/m^2.

The efficiency variation with Re for solar air heaters with smooth surfaces or no baffles, discrete V-type baffles, and continuous V-type baffles for fixed insolation values being 900 W/m^2, 800 W/m^2, and 700 W/m^2 were studied. The figure above

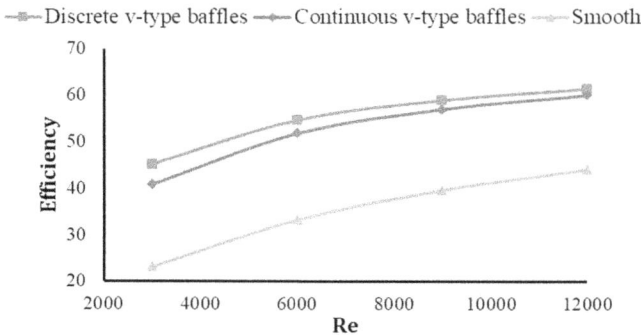

FIGURE 5.10 Efficiency variation with Re for discrete V-type baffles, continuous V-type baffles and without baffles SAH for insolation, I = 900 W/m^2.

shows the efficiency variation with *Re* for discrete V-type baffles, continuous V-type baffles, and without baffles solar air heater for insolation, I = 900 W/m². The minimum and maximum measured values of efficiencies are found between 45% and 61.5% for the discrete V-type baffles and 40.1% and 59.8% for the continuous V-type baffles, respectively. The increment of surface area and turbulence effect in between air and thus the absorber plate might happen when using baffles. It leads to an increasing amount of convective heat transfer rate. Thus, it improves the efficiency of the SAH. Efficiency rises with increasing *Re* for some time and then decreases for higher *Re* (beyond a certain value). Mechanical power through the variable speed electric blower is dominant. It is responsible for overcoming the frictional forces in the channel. The duct's increased friction losses are negligible at a lower Reynolds number, but due to increased turbulence, there is a noticeable increase in heat transfer from surfaces with baffles. Higher thermal efficiency is seen to correspond to roughness values that are higher. The conclusions established from the present study are in close agreement well with findings that were previously available [8].

The solar radiation ranges were 700 W/m², 800 W/m² & 900 W/m². There is a finding that the use of the periodic baffles of V-bent causes turbulence and increases in high-pressure drop with a considerable heat transfer increment. These discrete V-type baffles give the highest thermo-hydraulic performance parameter value being 3.12. The continuous V-type baffles induce the thermo-hydraulic performance parameter value being 2.22.

5.6   CONCLUSIONS

The use of the V-type baffles causes a substantial pressure to drop and provides significant heat transfer intensification. The value of Nu/N_{ss} is 5.36 for the discrete V-type solar air heater and, Nu/N_{ss} is 3.94 for the continuous V-type solar air heater.

Nusselt number ratio improves along with the rising value of *Re*. Using baffles compared to the smooth duct leads to a faster rate of heat transfer and supplies a greater thermal enhancement factor plus loss due to friction.

The average friction factor values are 6.97 for using the discrete V-type baffles and 6.12 for using the continuous V-type baffles, respectively. But the smooth plate supplies a lower friction factor and also provides a lower thermal enhancement factor.

This discrete V-type baffle provides the highest thermo-hydraulic performance parameter value of 3.12, while the continuous V-type solar air heater provides 2.21 respectively.

With an increasing value in Reynolds number, efficiency appears to increase for a period. For very high Reynolds numbers, it decreases.

The efficiency of the discrete V-type baffles is found to be higher than the continuous V-style and smooth plate SAH. The discrete V-type baffles have better efficiency than the continuous V-type and smooth plate SAHs. The fluctuation in thermal efficiency with different sun insolation values makes it known that when solar insolation is highest, efficiency is at its peak.

REFERENCES

[1] T. Alam, R. P. Saini, and J. S. Saini, "Experimentally investigation on heat transfer enhancement due to V-shaped perforated blocks in a rectangular duct of solar air heater," *Energy Convers. Management*, vol. 81, pp. 374–383, 2014.

[2] T. Rajaseenivasan, S. Srinivasan, and K. Srithar, "Comprehensive study on solar air heater with circular and V-type turbulators mounted on absorber plate," *Energy*, vol. 88, pp. 863–873, 2015.

[3] M. Bovand, S. Rashidi, and J. A. Esfahani, "Heat transfer enhancement and pressure drop penalty in porous solar heaters: numerical simulations," *Sol. Energy*, vol. 123, pp. 145–159, 2016.

[4] V. B. Gawande, A. S. Dhoble, D. B. Zodpe, and S. Chamoli, "Analytical approach for evaluation of thermo hydraulic performance of roughened solar air heater," *Case Stud. Therm. Eng.*, vol. 8, pp. 19–31, 2016.

[5] C. Nuntadusit, I. Piya, M. Wae-hayee, and S. Eiamsa-ard, "Heat transfer characteristics in a channel fitted with zigzag-cut baffles," *J. Mech. Sci. Technol.*, vol. 29, pp. 2547–2554, 2015.

[6] J. Hu, X. Sun, and Z. Li, "Numerical analysis of mechanical ventilation solar air collector with internal baffles," *Energy Build*, vol. 62, pp. 230–238, 2013.

[7] R. Ravi, and R. P. Saini, "Nusselt number and friction factor correlations for forced convective type counter flow solar air heater having discrete multi-V shaped and staggered rib roughness on both sides of the absorber plate," *Applied Thermal Engineering*, vol. 129, pp. 735–746, 2018.

[8] R. Kumar, R. Chauhan, M. Sethi, and A. Kumar, "Experimental study and correlation development for Nusselt number and friction factor for discretized broken V-pattern baffle solar air channel," *Experimental Thermal and Fluid Science*, vol. 81, pp. 56–57, 2017.

[9] A. Priyam, and P. Chand, "Thermal and thermohydraulic performance of wavy finned absorber solar air heater," *Solar Energy*, vol. 130, pp. 250–259, 2016.

[10] S. Skullong, P. Promvonge, C. Thianpong, N. Jayranaiwachira, and M. Pimsarn, "Heat transfer augmentation in a solar air heater channel with combined winglets and wavy grooves on absorber plate," *Applied Thermal Engineering*, vol. 122, pp. 268–284, 2017.

[11] Q. Li, C. Zheng, A. S. Jason, and A. T. Robert, "Design and analysis of a medium temperature, concentrated solar thermal collector for air-conditioning applications," *Applied Energy*, vol. 190, pp. 1159–1173, 2017.

[12] A. Ekadewi, Handoyo, D. Ichsani, and S. Prabowo, "Numerical studies on the effect of delta-shaped obstacles' spacing on the heat transfer and pressure drop in V-corrugated channel of solar air heater," *Sol. Energy*, vol. 131, pp. 47–60, 2016.

[13] S. Skullong, P. Promvonge, C. Thianpong, and M. Pimsarn, "Thermal performance in solar air heater channel with combined wavy-groove and perforated-delta wing vortex generators," *Appl. Therm. Eng.*, vol. 100, pp. 611–620, 2016.

[14] T. Tomar, G. Tiwari, and B. Norton, "Solar dryers for tropical food preservation: Thermophysics of crops, systems and components," *Solar Energy*, vol. 154, pp. 2–13, 2017.

[15] V. S. Hans, R. P. Saini, and J. S. Saini, "Heat transfer and friction factor correlations for a solar air heater duct roughened artificially with multiple v-ribs," *Solar Energy*, vol. 84, pp. 898–911, 2010.

[16] A. E. Kabeel, A. Khalil, S. M. Shalaby, and M. E. Zayed, "Experimental investigation of thermal performance of flat and v-corrugated plate solar air heaters with and without PCM as thermal energy storage," *Energy Conversion and Management*, vol. 113, pp. 264–272, 2016.

[17] M. J. Lewis, "Optimizing the thermohydraulic performance of rough surfaces," *Int. J. Heat Mass Transfer*, vol. 18, pp. 1243–1248, 1975.

[18] S. J. Kline, and F. A. McClintock, "Describing uncertainties in single sample experiments," *Mech. Eng.*, vol. 75, pp. 3–8, 1953.

[19] S. Chamoli, and N. Thakur, "Performance study of solar air heater duct having absorber plate with V down perforated Baffles," *Songklanakarian J. Sci. Technol.*, vol. 36, no. 2, pp. 201–208, 2014.

[20] P. Dutta, P. P. Dutta, and P. Kalita, "Thermohydraulic investigation of different channel height on a corrugated heat exchanger," in: AIP Conf. Proc., 2019. 10.1063/1.5096502.

[21] P. P. Dutta, and A. Kumar, "Development and Performance Study of Solar Air Heater for Solar Drying Applications," in: *Sol. Dry. Technol.*, 2017: pp. 579–601. 10.1007/978-981-10-3833-4_21.

[22] P. P. Dutta, "Prospect of renewable thermal energy in black tea processing in Assam: an investigation for energy resources and technology," Ph.D. diss., Tezpur University, India, 2014. https://shodhganga.inflibnet.ac.in/handle/10603/37571.

[23] P. Dutta, P. P. Dutta, and P. Kalita, "Thermal performance studies for drying of Garciniapedunculata in a free convection corrugated type of solar dryer," *Renew. Energy*, vol. 163, pp. 599–612, 2021. 10.1016/j.renene.2020.08.118.

[24] V. Goel, V. S. Hans, S. Singh, R. Kumar, S. K. Pathak, M. Singla, S. Bhattacharyya, E. Almatrafi, R. S. Gill, and R. P. Saini, *A comprehensive study on the progressive development and applications of solar air heaters.* Elsevier, 2021.

[25] H. S. Arunkumar, S. Kumar, and K. V. Karanth, "Analysis of a Solar Air Heater for augmented thermohydraulic 1 performance using 2 helicoidal spring shaped fins-A numerical study," *Renewable Energy*, vol. 960, pp. 31015–31016, 2020.

[26] A. R. Kalash, S. S. Shijer, and L. J. Habeeb, "Thermal Performance Improvement of Double Pass Solar Air Heater," *Journal of Mechanical Engineering Research and Developments*, vol. 43, pp. 355–372, 2020.

[27] C. P. Vora, V. H. Oza, and N. M. Bhatt, "Review of Passive Heat Transfer Augmentation Techniques for Solar Air Heaters," *Journal of Critical Reviews*, vol. 8405, pp. 2394–5125, 2020.

[28] A. Khanlari, A. Sözen, C. Şirin, A. D. Tuncer, and A. Gungor, "Performance enhancement of a greenhouse dryer: analysis of a cost-effective alternative solar air heater," *Journal of Cleaner Production*, vol. S0959-6526, no. 19, pp. 34542-1, 2019.

[29] A. K. Singh, N. Agarwal, and A. Saxena, "Effect of extended geometry filled with and without phase change material on the thermal performance of solar air heater," *Journal of Energy Storage*, vol. 39, p. 102627, 2021.

[30] H. Parsa, M. Saffar-Avval, and M. R. Hajmohammadi, "3D simulation and parametric optimization of a solar air heater with a novel staggered cuboid baffles," *International Journal of Mechanical Sciences, Elsevier*, vol. 205, p. 106607, 2021.

[31] J. D. D. Niyonteze, F. Zou, G. N. O. Asemota, W. Nsengiyumva, N. Hagumimana, L. Huang, A. Nduwamungu, and S. Bimenyimana, "Applications of Metaheuristic Algorithms in Solar Air Heater Optimization: A Review of Recent Trends and Future Prospects," *International Journal of Photoenergy*, vol. 36, p. 6672579, 2021.

[32] R. Khatri, S. Goswami, M. Anas, S. Sharma, S. Agarwal, and S. Aggarwal, "Performance evaluation of an arched plate solar air heater with porous aluminum wire mesh cylindrical fins," *Energy Reports*, vol. 6, pp. 627–633, 2020.

[33] S. Singh, "Experimental and Numerical Investigations of a Single and Double Pass Porous Serpentine Wavy Wiremesh Packed Bed Solar Air Heater," *Renewable Energy*, pp. 30977- 2, 2019.

[34] S. P. Shetty, N. Madhwesh, and K. V. Karanth, "Numerical analysis of a solar air heater with circular perforated absorber plate," *Solar Energy*, vol. 215: pp. 416–433, 2021.

[35] A. L. Antony, S. P. Shetty, N. Madhwesh, N. Y. Sharma, and K. V. Karanth, "Influence of stepped cylindrical turbulence generators on the thermal enhancement factor of a flat plate solar air heater," *Solar Energy*, vol. 198, pp. 295–310, 2020.

6 Design, Fabrication, and Testing of FSW Fixture for Conventional Milling Machine

Durga Rao Boddepalli, Siddhant Palai,
Rashmi Rekha Sahoo, Sritam Pattnayak, and
Anjan Kumar Mishra
Parala Maharaja Engineering College, Odisha, India

CONTENTS

DOI: 10.1201/9781003242291-6

6.1   INTRODUCTION

In THE WELDING INSTITUTE of the UK (TWI) in 1991, FSW was invented as the solid-state metal joining process [1]. FSW is the derivative of conventional friction welding which provides good-quality butt and lap joints. The material which is difficult to weld by conventional fiction welding can easily be welded using the FSW process. FSW simply provides defect-free welds in materials with poor fusion weldability [2]. It has some special benefits over traditional welding processes like fine microstructures, elimination of filler materials, absence of cracks in weld zones, reduce energy requirement, and many more other features which make it more convenient.

Generally, FSW welding is carried out in an FSW machine but we can also use a conventional milling machine with a suitable setup. For this setup, a fixture is used instead of vice to restrict the movement of the specimen in any direction. A fixture was created by S Budin et al. (2019) for the friction stir welding process by a conventional milling machine with the design and evaluation of different fabrication facilities to test its feasibility for Aluminum sheets [3]. For the FSW process to be carried out, the FSW tool is required whose material selection and geometry have major influences on the mechanical and microstructural properties of weld samples.

Friction Stir Welding is a superior alternative for welding to eliminate the creation of intermetallic compounds (IMCs) because of its solid-state mixing properties by modulating the heat generated with the help of a non-consumable tool consisting of 2 major components; shoulder and probe [4]. There are mainly three types of FSW tools: Fixed type, self-reacting, and adjustable type tool, where commonly used tool materials are H13 steel, High strength steel, SKD51 steel, and SKD61 steel. The various FSW tool shoulder and pin configurations are cylindrically threaded, three faces threaded, triangular, trivex, threaded conical, tri-flute, flared tri-flute, skew, etc. The manufacturing of different materials by FSW consists of Al alloys, Cu alloys, Mg alloys, Ti alloys, steels, Ni alloys, Polymers, and metal matrix composite.

The basic working of Friction Stir welding is very much simple and also cost-effective. A non-consumable rotating tool having the pin and specially fabricated shoulder plunged inside a workpiece until the shoulder touches the workpiece at the welding line and traverse along the welding line generates sufficient heat and plastic deformation of the material. Comparing the tool rotational and tool traveling direction, two sides arise: the advancing side and the retreating side. The various process parameters comprise rotational speed, transverse speed, tilt angle, tool offset, and downward force. Weld samples undergo various mechanical testing to obtain a good weld sample. The various mechanical testing includes tensile testing, Vickers hardness, NDT, etc.

6.2 METHODOLOGY

6.2.1 Fixture

During welding, we require zero movements to the workpieces in all directions. So, we have to design a fixture. Clamping of the workpiece, restriction of movements, and flexibility are most important for the design criteria. In machine tools, a fixture is a workpiece-locating and clamping device. Fixture is employed in welding and assembly inspection. The cutting tool is attached to the arbor of the conventional milling machine. The burden for accuracy changes from the user to the machine tool's construction when utilizing a fixture. The machining table can be attached instead of the fixture when a small number of parts are to be machined. On the other hand, a fixture is typically employed to store and locate work when the quantity of parts is significant enough to warrant its cost.

6.2.1.1 General Design Principles

a. The chief frame of the fixture must be sturdy enough to ensure the minimum deflection in the fixture. Cutting forces, workpiece clamping forces, and clamping to the machine table all contribute to fixture deflection. The fixture's mainframe should be heavy enough to prevent vibration and chatter.

b. Simple components can be used to construct frames, which can then be screwed or welded together as needed. Welding is an option for those elements of the frame that will be attached to the fixture permanently. Screws can be used to hold pieces that need to be changed frequently. When the body of the fixture has a complex shape, it can be cast from high-quality cast iron.

c. Clamping should be quick and involve little effort.

d. Clamps should be placed in such a way that they are conveniently accessible and removable.

e. Clamps should be supported by springs whenever possible so that clamps are held against the bolt head

f. If the clamp is to swing away from the work, it should be allowed to swing as far as is necessary for the workpiece to be removed.

g. All locator clamps should be visible and accessible to the operator for cleaning, positioning, and tightening.

h. Provision should be made for convenient chip disposal, so that chip storage does not obstruct the operation and chip removal during the operation does not obstruct the cutting process.

i. All wrench-adjustable clamps and support points should be of the same size. The fronts of the fixture should be able to operate all clamps and movable support points.

j. When the workpiece is placed in the fixture it should be stable. For a rough workpiece, three fixed points are needed. For a smooth workpiece, more than three fixed points may be needed. The placing of support points must be as wide as possible practically.

k. The workpiece's center of gravity should be surrounded by the three support points.
l. Without causing damage to the workpiece, the contact surface area of the support should be as little as feasible. The clamping of labor forces has caused this damage.
m. If support points or other components break, they are designed to be easily replaced.

6.2.1.2 Troubles in FSW

Proper fixture planning can help to dissipate heat from the job, resulting in better weld quality and performance [5]. A fixture was created to hold both the backplate and the workpiece [6]. An adaptable fixture was devised to grip the job being welded, considering the gap development and transverse movement of the workpiece [7]. The fixture needed to efficiently carry out the FSW operation in a traditional milling machine is not thoroughly explored in the literature. Friction stir welding is a technique that has been used to create custom equipment or to modify existing machine tool technology. FSW has a bright future since it will open up many cost-effective options in a variety of industries, including railway, maritime, transportation, refrigeration, and electrical [8].

Drawbacks of the FSW process: -

1. Special fixture is needed.
2. A visible hole has been created in the workpiece.
3. Initial or set up cost is high.
4. It is more rigid compared to the arc welding process.
5. It cannot use filler material.

6.2.1.3 Proposed Research Plan

The goal is to create an FSW setup and construct a defect-free weld. Any machine that can do a specific duty has long been regarded as crucial in the industry. Specialized machines are less accessible and usable due to their high cost. Here, machine adjustments are required to suit specific duties in a low-cost and straightforward manner. The exorbitant cost and lack of availability of the FSW machine prompted the creation of a new fixture. Throughout the friction stir welding process, various forces are also present. Torque, transverse, axial, and lateral forces are the four types of forces. The key problem was to endure all of these stresses during the friction stir welding process, and this was one of the goals of constructing the fixture.

6.2.1.4 Planning and Evolution of Fixture

The main objective of a friction stir welding fixture is to keep the work components in place while welding. Although, there is a scarcity of available information on the fixture design specifications. It is required to provide adequate area for the welding tool shoulder route, sufficient force to restrict workpiece movement, and enough heat sink to disperse welding heat in each joint design and fixture configuration.

Any solid body in free space has 12 degrees of freedom. So, the workpiece also has 12 degrees of freedom, which include six linear and six rotational movements along the coordinate axes. The 3-2-1 placement principle is the most widely employed. During a setup, some degrees of freedom must be restricted to find and align the workpiece concerning the friction stir welding tool. Because the true geometrical shape of supporting or restricting surfaces can vary, mainly on unfinished surfaces, the workpiece is preferred to be placed for the point supports. The vast surface of the backplate is used for the first three points, while the fixed side plate is used for the last two points. Instead of three pins, a flat base is used in this case. The movements are constrained in five directions when this is used. Instead of two pins, a flat supporting plate is utilized to restrict the three movements of the part around the Z-axis and in the direction of the Y-axis. A C-clamp has been used to limit the remaining four degrees of freedom.

A fixture is made up of a set of clamps and a locator. Locators find the location and orientation of a job, whereas clamps provide clamping pressures to keep the workpiece securely against the locator. Clamping was well-planned throughout the design of the machining fixture. The design is straightforward and inexpensive, and it can be mounted on any milling machine bed. A lot of components make up the fixture. Given the system's high cost and complexity, fabricating it in one piece would be too impossible. Many pieces were created independently in the beginning, then they were put together to make a single component. Parts are designed using the Dassault Systemes Solidworks Simulation 2016 software suite. This cuts down on time. Some flaws were detected in the fixture design. There were two notches present in the clamps which helped the clamps to slide on the backplate. But as the plates were fastened with nuts and bolts, there were chances of failure at the neck. Also, the two long nut bolts that hold the clamps limit the total tool travel distance.

6.2.1.5   Developed Fixture for Welding

Fundamental designs had been created to examine the effective design for the jigs and fixtures. The choice of the design is based on a decision-based method having five design considerations; minimum cost and manufacturability, the ability of heat dissipation, movement, and vibration stability. Following are the design changes that had been carried out in the old fixture.

a. The clamps were replaced by L clamps
b. The bolts connecting the clamps were removed as the L clamps are strong enough to be fastened with the backplate and bed.

6.2.2   FIXTURE DESIGN

The material which we had taken for the manufacturing of the fixture has a low thermal conductivity that would help in maintaining the temperature sufficiently high to facilitate plasticizing the material and its displacement. The fixture comprises the following components namely, the backplate, pressure bar, and L-clamp (Figure 6.1). The fixture specifications are mentioned in Table 6.1.

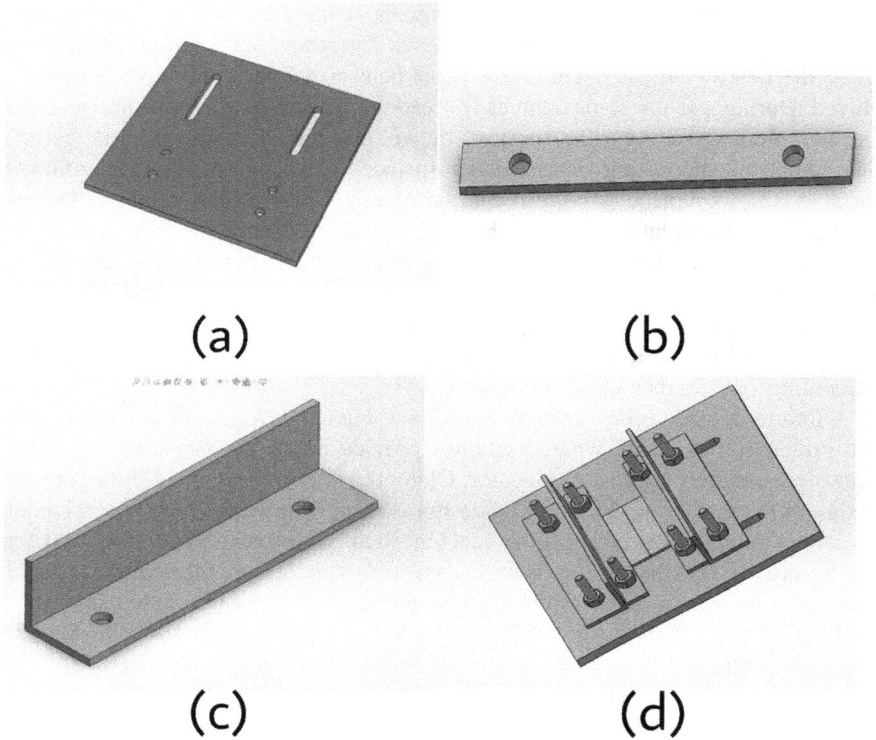

FIGURE 6.1   Fixture Components (a) Backplate (b) L Clamp (c) Pressure bar (d) Complete Fixture Assembly.

TABLE 6.1
Fixture Dimensions

Components	Dimensions in mm	Quantity	Material
L-Plate	220 × 50 × 50	2	Galvanized MS
Pressure Bar	220 × 25 × 5	2	Galvanized MS
Back Plate	300 × 300 × 11	1	Mild steel
Bolts and Nuts	Standard M10	8	Mild steel

6.2.3 DESIGN ANALYSIS

The fixture was analyzed in the Ansys software. The varying load was applied and the maximum stress and displacement values were obtained (Figure 6.2). The analysis of the fixture design was carried out and no observable deformation was observed. Hence the above fixture design was considered to be used in the FSW process. The fabrication process was carried out by following operations such as grinding, drilling, shaping, and milling.

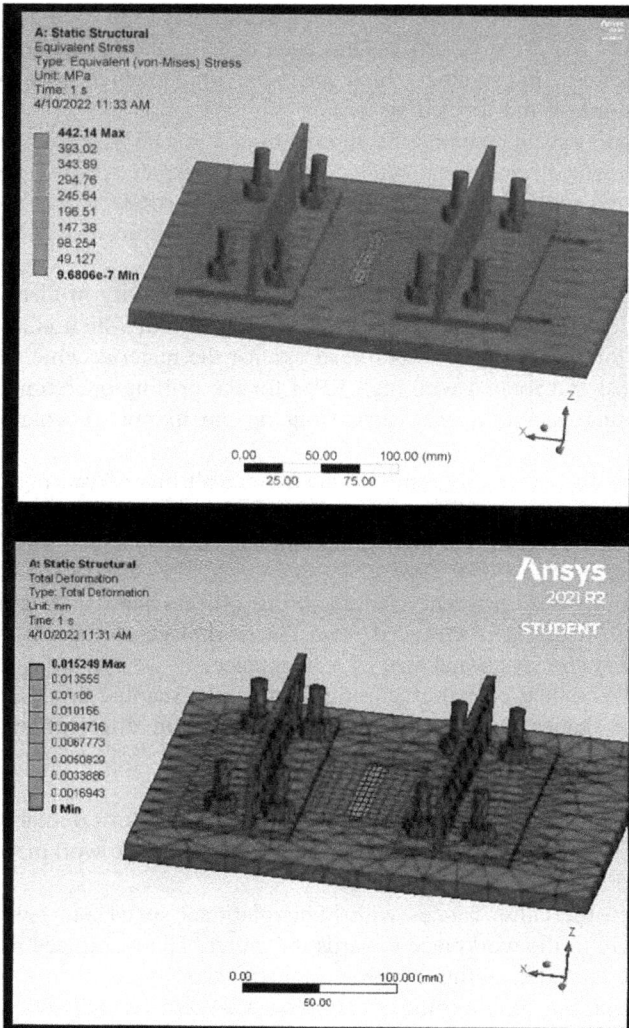

FIGURE 6.2 Fixture Analysis (a) Stress Analysis (b) Deformation Analysis.

6.2.4 FABRICATION OF FIXTURE

Following are the basic machining processes used for the manufacturing of fixtures: -

- Grinding: -
  - Grinding is a material removal process with the help of abrasives fused in the form of a rotating wheel. When the rotating abrasive wheel comes in contact with the workpiece, the abrasive particles work as small cutting tools and cut small chips from the workpiece.

- In the context of fixture fabrication, grinding is used for a good surface finish. The grinding process has been done on the backplate. The main objective for grinding which has been implemented was to provide frictionless and leveled surfaces.
- Grinding is an essential machining process for FSW fixtures because if the surface remains unlevelled then the workpieces which were to be welded will tilt at a certain angle in the portions where there is an uneven surface and proper welding of the workpieces will not occur.
- Drilling: -
  - Drilling is a machining process of creating a cavity of circular cross-section in a material by the application of a tool which is known as a Drillbit. The drill bit was forced against the material which was to be drilled and rotated with high RPM for the drilling operation.
  - Drilling operation was carried out on the fixture for connecting the bolts.
  - According to the design specifications, drill bits of varying diameters are used.
  - For threading of bits into it, tapping has been done.
- Shaping: -
  - Shaping is a different manufacturing process similar to planning in which material is removed from the workpiece surface. It is used to modify the shape and size of a workpiece.
  - In the context of manufacturing fixtures, the shaping process was used on a shaper machine for creating grooves in different positions at particular distances.
- Milling: -
  - Milling is a machining process in which cutters of circular shape are used to remove unwanted material by translating the workpiece into the cutter.
  - When the cutter rotates with high rotational speed and feed is given slowly to the workpiece towards the cutter. The workpiece passes past the cutter and milling of the workpiece takes place.
  - Concerning fixture milling operation was used on the workpiece with the help of a high-speed cutting tool for obtaining sharp and perpendicular edges.
  - The process is to be implemented with a horizontal milling machine.

After fabrication, the complete setup of the fixture is shown in Figure 6.3.

6.3 TOOL

To perform FSW, a friction stir welding tool is necessary which comprises a round shoulder and pin. Various factors are needed to design an FSW tool such as tool materials, pin length, shoulder diameter, shoulder shape, pin shape, etc. The fixture specifications are mentioned in Table 6.2. Tool shoulders were designed to produce enough heat to plasticize material. The shoulder produced

FIGURE 6.3 **All the components fastened with the back plate** (Part names: 1. Backplate; 2. L-clamp; 3. Pressure bar; 4. Workpieces to be placed; 5. Arbor; 6. Spindle; 7. Table; 8. Head; 9. Column; 10. Power feed).

TABLE 6.2
Tool Specification

Tool Parameters	Dimensions in mm
Tool material	H13 tool steel
Shoulder diameter	16
Pin length (plunge depth)	2.5
Tapered pin major diameter	4
Tapered pin minor diameter	3
Concave domed diameter	13
Concave angle	15
Total tool length	70

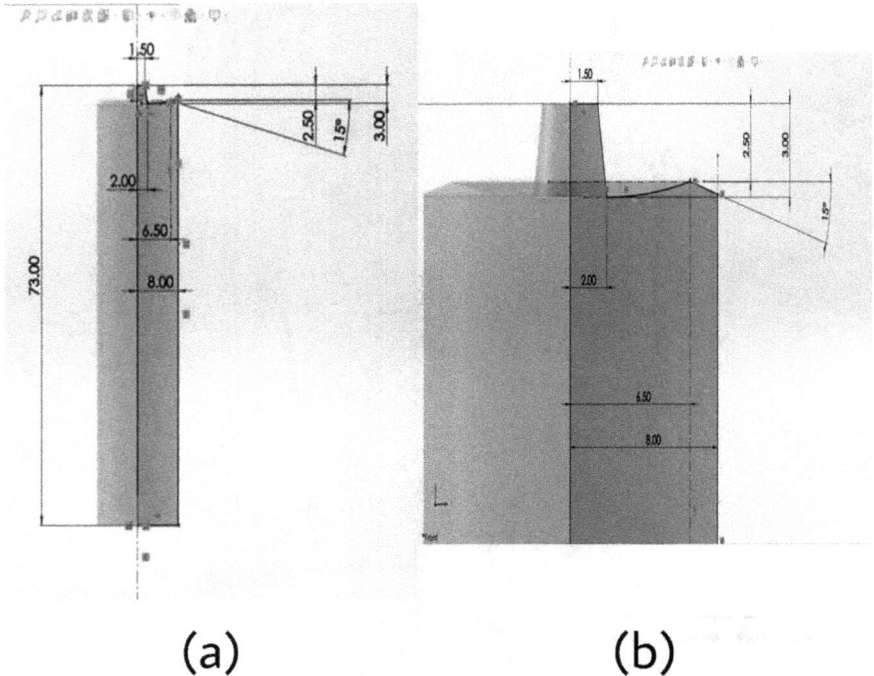

(a)   (b)

FIGURE 6.4   Tool Design (a) Side view of tool (b) Enlarged View of pin.

the downward forging action necessary for weld consolidation. During the tool plunge stage, material displaced by the pin was fed into the cavity within the tool shoulder.

Threads on pins were considered for the movement of material from the shoulder to the bottom of the pin. Forward movement of the tool forces new material into the cavity of the shoulder, pushing the existing material into the flow of the pin. The diameter of the tool's shoulder is proportional to the torque at a constant rotational speed. As the tool's shoulder diameter increases, the torque also increases. The pin length is slightly shorter than the thickness of the workpiece and its diameter is slightly larger than the thickness of the workpiece (Figure 6.4).

The analysis of the tool design was carried out and the developed design was then tested for any flaws in the software. The deformation observed in the tool was negligible and was approved for manufacturing. The machining process was carried out in the CNC lathe machine. Then the processed product was sent for grinding for better precision. After that, the tool underwent heat treatment in a furnace at 600°C for 48 h. The finished product then goes through oil hardening for longer tool life.

Tool material: -
For producing FSW welds of high quality, it is required that tool material should be selected properly. According to Meilinger and Torok [9] and Zhang et al. [10]

characteristics of the material which should be taken into consideration before selecting it for FSW: -

- No harmful reaction with the workpieces.
- Creep resistance at different temperatures, dimensional stability, and good strength of the material
- Good machinability for manufacturing complex structures on the shoulder and pin.
- Good wear resistance
- Doesn't have a high coefficient of thermal expansion.

There are various tool materials used for the FSW process. Some of the materials used as FSW tools are High-speed steel (HSS), Ni-alloys, ceramics, H-13 steel, etc.

In this paper, we have used H-13 steel for the FSW process. The main reason why we have selected H-13 steel is that it can be easily available, and has good machinability and thermal fatigue strength. They are also effective with both similar and dissimilar welding of metals for both butt and lap joints. It also has resistance to shock, abrasion, and deformation [11].

6.4 MATERIAL SELECTION

In this paper, A6061-T6 aluminum alloy and pure copper had been taken into consideration. Aluminum has a lightweight structure requiring a high strength-to-weight ratio and good corrosion resistance. Similarly, copper-based alloys have high strength, good ductility, and corrosion resistance. We have used a combination of Al-Al and Al-Cu weld samples.

6.5 PROCESS PARAMETER

For attaining the best quality FSW, the effect of process parameters like shoulder diameter, rotational speed, welding speed, tilting angle, and plunge depth should be identified. Researchers completed a series of tests to understand the influence of FSW parameters on weld appearance, microstructure, mechanical, electrical, and thermal properties [12]. FSW includes material movement and plastic deformation which are affected by welding parameters and joint design significantly.

In our project butt joint was taken into consideration. At first two metal pieces of even thickness were cut and placed on the backing plate which is kept on the machine bed. Those pieces were then clamped tightly to butting pieces from separation. After clamping the workpiece, the tool was then rotated and slowly brought down to a point where the tool tip touches the butting pieces. During the initial plunge of the toll the forces were fairly large and extra care was required to ensure that plates in the butt configuration don't separate [13]. After the plunge of the rotating tool, the tool then moves in a transverse direction along the joint line. When the surface of the tool is in contact with the surface producing weld along the abutting line. Tool rotation, tool transverse speed, plunge depth, tool offset, tilting angle, workpiece location, and workpiece thickness are some welding factors that are responsible for the performance and quality of the welded joints in the case of FSW. We have taken the tool rotation rate, tool transverse speed, and plunge depth to check the weld quality. Selection of process parameters for welding dissimilar materials like copper and aluminum plays a vital role in providing enough heat, softening the workpiece, and proper flowability in the stir zone, resulting in a defect-free, good-quality joint. For aluminum and copper FSW following are the important process parameters.

6.5.1 FSW Tool for Al-Cu Joining

Tool design and configuration are just as significant as other process elements in the production of a superior joint. During the welding operation, the FSW tool, which contains a pin and shoulder, affects the generation of heat and flow of material that further affects joint characteristics [14,15]. Proper design of the FSW tool improves joint performance and the Al-Cu FSW joint's performance. As a result, the tool material and its intrinsic qualities are critical for the FSW of dissimilar materials.

6.5.1.1 Materials for Al and Cu FSW Tools

The tool's geometry and properties must remain constant during the FSW process. The quality of joints depends upon the tool materials' wear resistance, hot hardness, sufficient density, machinability, the reactivity of the tool, microstructural homogeneity, and fracture hardness [16,17]. The MP159 cobalt-base super-alloy probe is made of AISI tool steel, high-strength steel, SKD51 steel, SKD61 steel, and H13 steel shoulder is an example of tool material. Several investigations have succeeded in executing welding on Al-Cu with heat-treated high-speed steel and tool steels with a hardness between 45 HRC to 62 HRC [18,19]. Tool steel used by researchers for Al-Cu FSW deteriorates and wears out at a higher rate as a result of the intense abrasion between Al and Cu. The primary causes of tool steel deterioration are heat instability and wear acceptance at increasing rotating speeds [20,21]. Due to their large surface area, threads and flutes on probes of various tool designs increase material flow behavior during FSW while simultaneously increasing heat input. Fixed, adjustable, and self-reacting or bobbin tools are the three categories of FSW tools. For workpieces of constant thickness, a fixed-type FSW tool is appropriate. An adjustable-type tool with two distinct components (pin and shoulder) is preferred since the pin length varies during welding. The bobbin-type tool, also known as the self-reacting type tool, is made up of three different bodies: bottom shoulder, top shoulder, and pin. The choice of tool material depends on the material's properties to be connected and its thickness [22].

6.5.1.2 FSW Tool Geometry

Since, the FSW tool influences the quantity of heat created, the plasticized material flow, force, and twisting moment encountered, selecting an optimal tool design is critical to achieving a better joint [23]. The diameter of the tool, surface profile, shoulder geometry, and pin geometry are all important features [24]. The tool shoulder has two major effects: (i) vertical compressive force and (ii) friction-induced thermal heat buildup. According to the research, tool shoulder diameter and surface features play a significant role in the frictional heat generated by rubbing action between the material and the shoulder, which influences weld qualities and defect formation [25–27]. Zhou et al. used three different pin profiles to evaluate the mechanical properties of an Al-Cu connection: pins with thread, featureless, and threaded with flutes [28]. The pin with threads has a 4.3 kN load, the highest among the three, trailed by the fluted threaded, and the featureless. Various investigations have revealed that maintaining the difference between workpiece thickness and probe length (Δt) minimal (Δt 0.5 millimeter) can result in a good fabrication along with the plate thickness. A large t may induce early weld failure due to insufficient material intermixing at the weld root. The shoulder-pin diameter ratio which is linked to sample thickness during dissimilar material welding is another factor to consider. It is slightly lower for FSW linking similar materials, while for different materials, it is significantly higher.

6.5.2 Effect of Welding Speed on Al-Cu FSW

The motion of the tool in the direction of the weld line is referred to as welding speed. During the welding process, it is one of the most critical factors that influence material flow and workpiece heating. Welding speed has an indirect effect on the quantity of heat input. A low quantity of heat is created due to high welding speed and vice versa.

Muthu and Jayabalan investigated the microstructural effects of Al-Cu joints with varying welding speeds [29]. This research found that increasing the welding speed from 50 to 80 mm/min improved the tensile strength of Al-Cu joints. The amount of heat input and the creation of IMCs are the key reasons for this. Because higher welding speeds result in reduced heat input and fewer IMCs, the stir zone (SZ) becomes less brittle. Welding speeds greater than 80 mm/min, on the other hand, resulted in low joint strength.

Another study discovered that faster welding speeds produced less heat during the operation, resulting in an incorrect mixture of Al-Cu. Due to the presence of flaws such as cavities and voids, weld joints are partial and of poor quality [30]. To investigate the effects of different welding speeds on tensile shear strength, Saied et al. performed friction stir lap welding on AA1060 aluminum alloy and pure copper [31]. The failure load grew with increasing weld speed until it reached a maximum value, after which it began to decrease. This was due to the low welding speed, which caused significant microcrack formation at Al-Cu contacts, reducing fracture strength.

6.5.3 Influence of Tool Rotating Speed on Al-Cu FSW

The reaction temperature depends upon the tool's rotation speed during dissimilar FSW. The tool's axis rotation affects frictional heat generation and delivers proper axial thrust, which influences the flow of material, stir zone size, and most importantly, the creation of IMCs. In most circumstances, the heat input is proportional to the spinning speed. As rotational speed increases, the stressed region grows larger, and the highest strain zone moves from the weld's initial retreating side (RS) to the advancing side (AS). This means that the tool's rotational speed affects the position of the weld fracture. Xue et al. used FSW on copper and 1060 aluminum alloy at various RPM and discovered that the lower the rotational speed, the higher weld defects such as porosity in the weld were created [18]. Copper incorporation into the aluminum matrix was challenging due to the less reaction temperature within the weld zone, resulting in pores. During FSW, the reduced rotation speed, in contrast to the welding speed, results in less heat input and an insufficient reaction temperature. As a result, incorrect plastic deformation occurs inside the stir zone, leading to an increase in weld defects such as voids and macrocracks. Higher RPM increases the temperature at the interface, lowering the flow stress of metal and facilitating material blending of aluminum and copper [32].

Bisadi et al. noticed weld defects like voids and microcracks at low speeds in the Al-Cu weld interface, which diminished as the rotational speed increased [33]. Softening of materials was achievable due to the increase in welding temperature generated by high RPM, allowing for homogeneous mixing of aluminum and copper. Due to the compounding effect of these process parameters, many studies have coined the term revolutionary pitch (v/ω), which affects weld quality. A higher revolutionary pitch value indicates fast welding (cold welding), while a lower revolutionary pitch value indicates sluggish welding (hot welding). Insufficient material flow and low peak temperature come from increased pitch ratios, whereas too low a value results in poor material flow and higher liquidation, which encourages IMC growth, both of which are detrimental to weld quality. Lower rotating speed causes low-temperature production in the nugget zone (NZ), and the volume percent of the coarse second phase (strengthening) particles has increased. Second-phase particles fracture more at higher rotating speeds, resulting in particle separation in other sections of the TMAZ. Due to the fact that the temperature range, which dictates the parameters, can result in weak joint characteristics, more research is needed to determine the best variation of RPM and welding speeds for Al-Cu FSW.

6.5.4 Effect of Plunge Depth on Al-Cu FSW

In FSW [34,35], one of the important aspects for securing better joint quality is the plunge depth of the workpiece. The effect of insertion depth on the shear strength of magnesium alloy lap joints using FSW was investigated, and it was discovered that increasing the tool pin's length and penetration depth into the bottom workpiece enhanced the joint's shear strength [36]. The increased plunge depth supports variations of mechanical interlocking mechanisms and forms a better metallic bond of the top and bottom workpiece in lap joint configuration due to the stirring effect of the tool pin. Using two different pin length tools, Elrefaey et al. took 2 mm thick aluminum and 1 mm thick copper as a lap joint to examine the influence of pin insertion on Al-Cu weld strength [37]. Whenever the pin insertion is 2 mm, with no insertion to the copper sheet the joint is weak whereas when the pin insertion is 2.1 mm and 0.1 mm into copper the joint shows improved strength with optimal rotational and welding speed. Copper shards of various sizes and structures were strewn throughout the aluminum matrix randomly. This indicates that the two metals have been mechanically combined. This enhances the performance of the joint by promoting strong Al-Cu contact bonding [4]. Guan et al. achieved successful Al-Cu junctions by creating Al-Cu FSW lap joints at high rotational speeds with no-tool insertion to lower copper workpiece [32]. Due to proper mixing at higher RPM the joint strength is enhanced.

6.5.5 EFFECT OF TOOL TILT ANGLE ON AL-CU FSW

The angle between the FSW tool and the perpendicular surface to the workpiece is called the tilt angle. The tool has a 0° or no tilt angle when orthogonal to the workpiece surface. During the friction stir welding of various materials, it is crucial to employ the ideal tilt angle since it keeps the softened material below the tool's shoulder and enhances the pressing action, allowing for maintaining consistent material flow through the moving of the plasticized material [38].

Mehta et al. employed FSW to join aluminum and copper and adjusted the tool angle with respect to the plate to examine the impact of the tool tilt angle on Al-Cu weld characteristics [39]. The flash defect was reduced when the tilt angle was increased from 0° to 4°. Furthermore, a larger tilt angle applied more compressive force beneath the shoulder area, allowing the tool to forge the agitated material downwards and prevent it from spreading outside the shoulder diameter, resulting in a strong connection and enhanced weld aesthetics. The hardness of the Al-Cu joint area was also improved by a larger tilt angle.

Another experiment with varied friction stir lap welding of aluminum and copper found that when tool tilt angles of 2°–4° were assumed, the maximum failure load of Al-Cu joints was obtained [40]. The previous study has shown that the tool tilt angle affects the joint's strength and integrity. As a result, to produce the optimum weld outcomes, an optimal tilt angle must be used [41]. The various process parameters are mentioned in Table 6.3.

TABLE 6.3
Various Process Parameters

Sl No.	Feed (mm/min)	Spindle (rpm)	Tool Type	Depth of Cut	Workpiece
1.	16	2000	Plane Tapered	2.8	Al+Al
2.	60	2000	Threaded Tapered	3.0	Al+Al
3.	32	2000	Plane Tapered	2.8	Cu+Al
4.	16	570	Plane Tapered	2.8	Cu+Al
5.	25	2000	Threaded Tapered	2.8	Al+Al
6.	32	570	Threaded Tapered	2.8	Al+Al
7.	40	570	Threaded Tapered	2.8	Al+Al
8.	32	2000	Plane Tapered	2.8	Al+Al
9.	32	2000	Threaded Tapered	2.8	Al+Cu
10.	32	2000	Threaded Tapered	2.8	Cu+Cu

6.6 TESTING

After the FSW process, Al-Cu and Al-Al alloys show different mechanical as well as chemical properties which are different from their parent properties. In order to identify those new properties, two tests had been conducted to determine the entity value in our post-experiment phase these testing experiment machines are listed below

- Universal testing machine
- Micro Vickers hardness tester

6.6.1 UNIVERSAL TESTING MACHINE

A universal testing machine (UTM), also known as the universal tester is used to examine the tensile and compressive strength of the material. In this test, the pre-weld alloy was tested to calculate the tensile strength. So, it was cut into standard size to fit in the machine properly. Then the machine was operated and the test was done. After the milling was performed the filing operation was carried out for the production of tensile strength specimens. Then the material was gripped at both ends by an apparatus and pulled lengthwise until it got fractured, due to that pulling load material changed by its length or displacement. The load was then converted to a stress value and the displacement was converted to a strain value.

6.6.2 MICRO VICKERS HARDNESS TESTER

The Vickers hardness test is used to measure the hardness of the material. The basic principle is to observe the material's ability to resist plastic deformation from a standard source. The unit of hardness is given by the Vickers pyramid number (HV) or Diamond pyramid hardness (DPH) [42]. The specimen for the hardness test was prepared with the standard dimension. After preparing the specimen the surface was polished using sandpaper, and then it was placed on an anvil. The anvil was then lifted to the optical head until we had a clear vision of the surface under the microscope. Using a manual turret, we substituted the optical head with Vickers and Knoop indenter which is the diamond indenter. Then the indentation was done with the load of 0.5 kg for 10 sec on the weld zone and at a distance of 30 mm on both sides. After indenting, we determined the two diagonals d1 and d2, and the area of the sloping surface of the indentation was calculated. After the diagonal is obtained the Vickers test machine is automatically calibrated and the value for hardness was given.

The Vickers hardness is the quotient obtained by dividing the kgf load by the square mm area of indentation.

F = load in kgf

d = arithmetic means of the two diagonals, d1 and d2 in mm

HV = Vickers hardness

$$HV = \frac{2F sin\frac{136°}{2}}{d^2}$$

$$HV = 1.854\frac{F}{d^2}$$

6.7 MICROSTRUCTURAL ANALYSIS OF AL-CU FSW

The combination is aggressively swirled, resulting in four different zones: the stir zone (SZ), the thermomechanically affected zone (TMAZ), the heat affected zone (HAZ), and the base metal zone (BZ) (BM). Frictional heat causes the workpiece in contact with the pin and shoulder at the stir zone to reach the maximum temperature and strain energy. The grain size in the SZ is very small compared to the base metals because of dynamic recrystallization. Based on the process parameters, the grain size was reduced up to 96 percent, giving the SZ the highest microhardness and strength. In some cases, at SZ an onion ring is formed of a quasi-spiral shape. According to Tongue et al., for the formation of varying layers of peak and moderate strain rate, material flow caused by the rotation and translation of tools is responsible [43].

TMAZ is the next closest zone to the stir zone and some heat flows from SZ to TMAZ which causes slight grain deformation. The lack of plastic strain, on the contrary, minimizes grain size deformation when compared to the SZ. The material in TMAZ is discharged away from the material flow direction. A flash fault, or excessive outward material flow, shows that the welding conditions were not appropriate. The grain structure in HAZ is dependent upon the temperature adhered to residual heat. The grain structure in the BM, on the other hand, remains unchanged since no heat or strain energy is present in the zone to cause deformation. Narasimharaju and Sankunny [44] compared welding RPM and grain size at the stir zone and found that at higher RPM the grain size is larger and at lower RPM the grain size is smaller. The grain size in SZ has an inverse relationship with microhardness. The microstructural analysis obtained for Al-Cu FSW is shown in Figure 6.5.

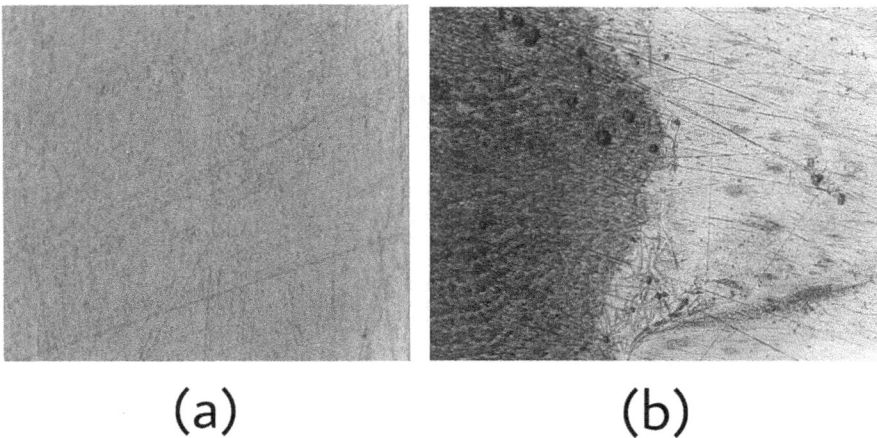

(a)   (b)

FIGURE 6.5 Microstructure (a) Al-Al 50X threaded pin profile (b) Al-Cu 10X threaded pin profile.

TABLE 6.4
Results of Micro Vickers Hardness

Sr. no.	Workpiece	Tool Feature	Tool Rotational Speed(rpm)	Transverse Speed (mm/min)	Micro Hardness (HV)
01	Al+Al	Threaded Tapered	2000	25	45.26
02	Al+Al	Plane Tapered	2000	32	26.08
03	Al+Cu	Threaded Tapered	2000	32	97.86
04	Cu+Cu	Threaded Tapered	2000	32	55.57

TABLE 6.5
Results of Tensile Test

Sr. no.	RPM	Feed (mm/min)	Tensile Strength (MPa)	Workpiece	Tool Profile
01	2000	16	43.6	Al+Al	Plane Tapered
02	2000	60	121.1	Al+Al	Threaded Tapered
03	2000	32	44.6	Cu+Al	Plane Tapered
04	570	16	61.8	Cu+Al	Plane Tapered
05	2000	25	126.8	Al+Al	Threaded Tapered
06	570	32	94	Al+Al	Threaded Tapered
07	570	40	94.3	Al+Al	Threaded Tapered

6.8   RESULTS

In this report, we studied the effect of FSW process parameters on the mechanical properties of friction stir welded 6061 T6 Aluminum alloy and copper. The hardness and tensile test results obtained from the test done on the weld sample for various parameters are tabulated in Table 6.4 and Table 6.5 respectively. The main parameters which affect the micro-hardness value are shoulder geometry and pin profile. The main role of these two elements of the tool is to create a downward force on material that comes outside due to the plunging of the pin and proper mixing of that material. Also, the proper design of the shoulder and the pin surface create an effective downward force and smooth surface finish, which is most important for the soundness of the weld.

6.9   DISCUSSION

- This paper aims to show a better way of welding two dissimilar materials (Aluminum 6061-T6 and pure copper) by selecting better process parameters to obtain the optimal properties of the weld by the FSW Process.
- Al6061-T6 alloy was successfully welded by FSW and no macro-level defect was found under the following range of process parameters; tool

material of H13, tool rotational speed from 570–2000 rpm, and welding speed of 16–60 mm/min.

- It is noted that the fixture that we have designed for FSW can withstand a maximum load of 179 kN due to which the maximum stress and deformation obtained are 442.14 Mpa and 0.0015259, respectively.
- It is noted that Friction Stir welding (FSW) leads to grain refinement caused due to dynamic recrystallization during the FSW and enhancement of other mechanical properties of alloy such as an increase in strength and increase in micro hardness also occurs.
- We observe that the hardness of the welded region is higher than the parent metal. We also observe the increase in strength with an increase in feed rate account with threaded tapered tool profile.
- It has been discovered that raising the rotating speed increases micro hardness, which then declines gradually. The improved hardness of the stir zone can be attributed to two factors. To begin with, grain refinement plays a crucial role in material strengthening since the grain size of the stir zone is much finer than that of the base metal. Second, the tiny particles of intermetallic compounds contribute to increased hardness. The micro hardness decreases as the rotating speed increases. This is due to excessive heat generation, which softens the material and produces a decrease in micro hardness. [24]
- It is also determined that when the tensile test was conducted the highest tensile strength of 126.8 MPa is observed while operating at 2000 rpm and feed of 25 mm/min with the threaded tapered tool profile. Under similar operating conditions, the second highest strength of 121.1 MPa operating at the same 2000 rpm and a different feed at 60 mm/min with the same threaded pin profile, and the lowest strength was observed with a plain tapered tool profile which was 43.6 MPa while operating at 2000 rpm with a feed of 16 mm/min. The overall result obtained from the tensile test was found to be satisfactory.
- It was observed from the above test that the best quality weld was achieved with threaded tapered tool profile as compared to plain tapered tool profile.

6.10 CONCLUSION

We have successfully designed and fabricated the FSW fixture which is used in conventional milling machines for welding materials such as Al-Al samples and Al-Cu samples with the help of different process parameters.

The fixture which we have designed, demonstrates great potential to fully utilize the existing conventional milling machine for FSW resulting in the quality weld. The welded samples were tested for different mechanical properties.

REFERENCES

[1] Anon, "Friction welding," *Welding Journal (Miami, Fla)*, vol. 78, no. 4, p. 56, 1999, doi: 10.31399/asm.hb.v06.a0001381.

[2] R. Nadda, M. Babal, N. Jalan, and C. K. Nirala, "Microfriction stir welding of AA 6061-T6 thin sheets using in-house developed fixture," *Journal of Micromanufacturing*, vol. 3, no. 1, pp. 5–12, 2020, doi: 10.1177/2516598419895837.

[3] S. Budin, N. C. Maideen, K. M. Hyie, and S. Sahudin, "Design and Development of Manufacturing Facilities for Friction Stir Welding Process using Conventional Milling Machine," *IOP Conference Series: Materials Science and Engineering*, vol. 505, no. 1, pp. 0–7, 2019, doi: 10.1088/1757-899X/505/1/012006.

[4] V. P. Singh, S. K. Patel, A. Ranjan, and B. Kuriachen, "Recent research progress in solid state friction-stir welding of aluminium–magnesium alloys: A critical review," *Journal of Materials Research and Technology*, vol. 9, no. 3, pp. 6217–6256, 2020, doi: 10.1016/j.jmrt.2020.01.008.

[5] P. Yogesh, C. Arun, R. Arunkumar, S. Muruganantham, and P. Abinesh, "Elimination of rejection in injector manufacturing process by using hydraulic fixture," *International Journal of Mechanical Engineering and Technology*, vol. 8, no. 3, pp. 331–343, 2017.

[6] Daniela Lohwasser, and Zhan Chen, *Friction stir welding: from basics to applications.* 2010.

[7] L. Fratini, F. Micari, and V. Ruisi, "A new fixture for FSW processes of titanium alloys," *CIRP Annals - Manufacturing Technology*, vol. 59, pp. 271–274, Dec. 2010, doi: 10.1016/j.cirp.2010.03.003.

[8] M. Peel, A. Steuwer, M. Preuss, and P. Withers, "Microstructure, mechanical properties and residual stresses as a function of welding speed in aluminium AA5083 friction stir welds," *Acta Materialia*, pp. 4791–4801, Sep. 2003, doi: 10. 1016/S1359-6454(03)00319-7.

[9] A. Meilinger and I. Torok, "THE IMPORTANCE OF FRICTION STIR WELDING TOOL Ákos Meilinger, Imre Török," *Production Processes and Systems*, vol. 6, no. 1, pp. 25–34, 2013.

[10] Y. N. Zhang, X. Cao, S. Larose, and P. Wanjara, "Review of tools for friction stir welding and processing," *Canadian Metallurgical Quarterly*, vol. 51, no. 3, pp. 250–261, 2012, doi: 10.1179/1879139512Y.0000000015.

[11] K. Chiteka, "Friction Stir Welding/Processing Tool Materials and Selection," *International Journal of Engineering Research and Technology (IJERT)*, vol. 2, no. 11, pp. 8–18, 2013.

[12] M. Verma, S. Ahmed, and P. Saha, "Challenges, process requisites/inputs, mechanics and weld performance of dissimilar micro-friction stir welding (dissimilar μFSW): A comprehensive review," *Journal of Manufacturing Processes*, vol. 68, pp. 249–276, Aug. 2021, doi: 10.1016/j.jmapro.2021.05.045.

[13] M. A. Siddiqui, S. a. H. Jafri, P. K. Bharti, and P. Kumar, "Friction Stir Welding as a Joining Process through Modified Conventional Milling Machine: A Review," *International Journal of Innovative Research & Development*, vol. 3, no. 7, pp. 149–153, 2014.

[14] A. Banik, A. Saha, J. Barma, U. Acharya, and C. Subhash, "Determination of best tool geometry for friction stir welding of AA 6061-T6 using hybrid PCA-TOPSIS optimization method," *Measurement*, vol. 173, Oct. 2020, doi: 10.1016/ j.measurement.2020.108573.

[15] J. Li, Y. Shen, W. Hou, and Y. Qi, "Friction stir welding of Ti-6Al-4V alloy: Friction tool, microstructure, and mechanical properties," *Journal of Manufacturing Processes*, vol. 58, pp. 344–354, Oct. 2020, doi: 10.1016/j.jmapro.2020.08.025.

[16] R. Rai, A. De, H. K. D. H. Bhadeshia, and T. DebRoy, "Review: friction stir welding tools," *Science and Technology of Welding and Joining*, vol. 16, no. 4, pp. 325–342, May 2011, doi: 10.1179/1362171811Y.0000000023.

[17] L. Cui, M. Maeda, and K. Nogi, "Effect of tool shape on mechanical properties and microstructure of friction stir welded aluminum alloys," *Materials Science and Engineering: A*, vol. 419, pp. 25–31, Mar. 2006, doi: 10.1016/j.msea.2005.11.045.

[18] P. Xue, D. R. Ni, D. Wang, B. Xiao, and Z. Y. Ma, "Effect of friction stir welding parameters on the microstructure and mechanical properties of the dissimilar Al–Cu joints," *Materials Science and Engineering A-structural Materials Properties Microstructure and Processing - MATER SCI ENG A-STRUCT MATER*, vol. 528, pp. 4683–4689, May 2011, doi: 10.1016/j.msea.2011.02.067.

[19] W. Hou *et al.*, "Enhancing metallurgical and mechanical properties of friction stir Butt welded joints of Al–Cu via cold sprayed Ni interlayer," *Materials Science and Engineering: A*, vol. 809, p. 140992, Feb. 2021, doi: 10.1016/j.msea.2021.140992.

[20] M. P., G. Reddy, and A. Sharma, *Evolution and Current Practices in Friction Stir Welding Tool Design*. 2020.

[21] P. Sahlot and A. Arora, "Numerical model for prediction of tool wear and worn-out pin profile during friction stir welding," *Wear*, vol. 408, May 2018, doi: 10.1016/j.wear.2018.05.007.

[22] Y. Du, T. Mukherjee, P. Mitra, and T. DebRoy, "Machine learning based hierarchy of causative variables for tool failure in friction stir welding," *Acta Materialia*, vol. 192, no. April, pp. 67–77, 2020, doi: 10.1016/j.actamat.2020.03.047.

[23] L. H. Shah, S. Walbridge, and A. Gerlich, "Tool eccentricity in friction stir welding: a comprehensive review," *Science and Technology of Welding and Joining*, vol. 24, no. 6, pp. 566–578, Aug. 2019, doi: 10.1080/13621718.2019.1573010.

[24] X. Wang, Y. Pan, and D. Lados, "Friction Stir Welding of Dissimilar Al/Al and Al/Non-Al Alloys: A Review," *Metallurgical and Materials Transactions B*, vol. 49, May 2018, doi: 10.1007/s11663-018-1290-z.

[25] L. Trueba, G. Heredia, D. Rybicki, and L. B. Johannes, "Effect of tool shoulder features on defects and tensile properties of friction stir welded aluminum 6061-T6," *Journal of Materials Processing Technology*, vol. 219, no. C, pp. 271–277, 2015, doi: 10.1016/j.jmatprotec.2014.12.027.

[26] Q. Chu, S. J. Hao, W. Y. Li, X. W. Yang, Y. F. Zou, and D. Wu, "Impact of shoulder morphology on macrostructural forming and the texture development during probeless friction stir spot welding," *Journal of Materials Research and Technology*, vol. 12, pp. 2042–2054, 2021, doi: 10.1016/j.jmrt.2021.04.013.

[27] K. K. Mugada and A. Kumar, "Effect of knurling shoulder design with polygonal pins on material flow and mechanical properties during friction stir welding of Al–Mg–Si alloy," *Transactions of Nonferrous Metals Society of China*, vol. 29, pp. 2281–2289, Nov. 2019, doi: 10.1016/S1003-6326(19)65134-4.

[28] L. Zhou, R. X. Zhang, G. H. Li, W. L. Zhou, Y. X. Huang, and X. G. Song, "Effect of pin profile on microstructure and mechanical properties of friction stir spot welded Al-Cu dissimilar metals," *Journal of Manufacturing Processes*, vol. 36, no. July, pp. 1–9, 2018, doi: 10.1016/j.jmapro.2018.09.017.

[29] M. F. X. Muthu and V. Jayabalan, "Tool travel speed effects on the microstructure of friction stir welded aluminum-copper joints," *Journal of Materials Processing Technology*, vol. 217, pp. 105–113, 2015, doi: 10.1016/j.jmatprotec.2014.11.007.

[30] Y. Fotouhi, S. Rasaee, A. Askari, and H. Bisadi, "Properties in dissimilar butt friction stir welding of al5083–copper sheets," *Engineering Solid Mechanics*, vol. 2, no. 3, pp. 239–246, 2014, doi: 10.5267/j.esm.2014.3.001.

[31] T. Saeid, A. Abdollah-zadeh, and B. Sazgari, "Weldability and mechanical properties of dissimilar aluminum-copper lap joints made by friction stir welding," *Journal of Alloys and Compounds*, vol. 490, no. 1–2, pp. 652–655, 2010, doi: 10.1016/j.jallcom.2009.10.127.

[32] Q. Guan, H. Zhang, H. Liu, Q. Gao, M. Gong, and F. Qu, "Structure-property characteristics of Al-Cu joint formed by high-rotation-speed friction stir lap welding without tool penetration into lower Cu sheet," *Journal of Manufacturing Processes*, vol. 57, no. May, pp. 363–369, 2020, doi: 10.1016/j.jmapro.2020.07.001.

[33] H. Bisadi, S. Rasaee, and Y. Fotoohi, "Studying of tool rotation speed on mechanical properties of copper-Al5083 butt joint welded by friction stir welding," *Proceedings of the Institution of Mechanical Engineers, Part B: Journal of Engineering Manufacture*, vol. 229, no. 10, pp. 1734–1741, 2015, doi: 10.1177/0954405414539491.

[34] Y. Wei, J. Li, J. Xiong, and F. Zhang, "Effect of tool pin insertion depth on friction stir lap welding of aluminum to stainless steel," *Journal of Materials Engineering and Performance*, vol. 22, no. 10, pp. 3005–3013, 2013, doi: 10.1007/s11665-013-0595-y.

[35] V. Chitturi, S. R. Pedapati, and M. Awang, "Effect of tilt angle and pin depth on dissimilar friction stir lap welded joints of aluminum and steel alloys," *Materials*, vol. 12, no. 23, pp. 1–11, 2019, doi: 10.3390/ma122333901.

[36] X. Cao and M. Jahazi, "Effect of tool rotational speed and probe length on lap joint quality of a friction stir welded magnesium alloy," *Materials and Design*, vol. 32, no. 1, pp. 1–11, 2011, doi: 10.1016/j.matdes.2010.06.048.

[37] A. Elrefaey, M. Takahashi, and K. Ikeuchi, "Microstructure of Aluminum/Copper Lap Joint by Friction Stir Welding and Its Performance," *Journal of High Temperature Society*, vol. 30, no. 5. pp. 286–292, 2004. doi: 10.7791/jhts.30.286.

[38] R. S. Mishra and Z. Y. Ma, "Friction stir welding and processing," *Materials Science and Engineering R: Reports*, vol. 50, no. 1–2, pp. 1–78, 2005, doi: 10.1016/j.mser.2005.07.001.

[39] K. P. Mehta and V. J. Badheka, "Effects of tilt angle on the properties of dissimilar friction stir welding copper to aluminum," *Materials and Manufacturing Processes*, vol. 31, no. 3, pp. 255–263, 2016, doi: 10.1080/10426914.2014.994754.

[40] M. Akbari, R. Abdi Behnagh, and A. Dadvand, "Effect of materials position on friction stir lap welding of Al to Cu," *Science and Technology of Welding and Joining*, vol. 17, no. 7, pp. 581–588, 2012, doi: 10.1179/1362171812Y.0000000049.

[41] M. S. M. Isa *et al.*, "Recent research progress in friction stir welding of aluminium and copper dissimilar joint: a review," *Journal of Materials Research and Technology*, vol. 15. Elsevier Editora Ltda, pp. 2735–2780, Nov. 01, 2021. doi: 10.1016/j.jmrt.2021.09.037.

[42] K. Chawla, *Mechanical behaviors of materials*. 1999.

[43] A. Tongne, M. Jahazi, E. Feulvarch, and C. Desrayaud, "Banded structures in friction stir welded Al alloys," *Journal of Materials Processing Technology*, vol. 221, no. March, pp. 269–278, 2015, doi: 10.1016/j.jmatprotec.2015.02.020.

[44] S. R. Narasimharaju and S. Sankunny, "Microstructure and fracture behavior of friction stir lap welding of dissimilar aa 6060-t5/ pure copper," *Engineering Solid Mechanics*, vol. 7, no. 3, pp. 217–228, 2019, doi: 10.5267/j.esm.2019.5.002.

7 Numerical Analysis of Fin Heat Transfer in Radiators Using Simulation Software Comsol Multiphysics 5.5

Shivashree Sharma and Saroj Yadav
Dibrugarh University Institute of Engineering and
Technology, Dibrugarh University

CONTENTS

DOI: 10.1201/9781003242291-7

7.1   INTRODUCTION

In thermal engineering, the intensification of heat transfer is a major topic to be discussed. The damaging effects caused by overheating or burning which should be avoided, the excessive heat generated from the system components should be removed and this acts as a crucial part. For the sake of maintaining acceptable temperature levels, for temperatures below 100°C, in electronic devices traditional cooling by air is used.

There are three ways in which heat can be transferred. That includes conduction, convection and radiation. In general, through modes of conduction and convection, the transfer of heat takes place from the surface, as the conductivity of a material is fixed and it can be intensified if the area of heat transfer of the surface can be maximized, or it can be done by both. The fins are extended surfaces which are generally used to amplify the heat transfer used in a lot of industries. These fins are used in numerous varied applications which include electronics, automobile radiators, refrigerators etc. This chapter is mainly concerned with radiators which are actually heat exchangers, which serve the purpose of transferring thermal energy between a source and a working fluid. The heat transfer rate depends on the heat transfer coefficient (h), effective surface area and the temperature difference between surfaces and the surrounding fluid. In convective mode, the value of (h) is the function of the properties of the surrounding fluid and the average velocity of the fluid over the surface. So, by increasing the surface area the overall dimension of the system will increase, thereby increasing the cost and size of the system.

Fins are usually employed to maximize the rate of transfer of heat by increasing the surface area. But, fins produce a conductive resistance to heat transfer from the original surface. The fin's entire surface area may not be utilized efficiently due to the geometric configuration of the system. These fins are used in varied applications which include automobile radiators, electronics, refrigerators, etc. This chapter is concerned with heat transfer in radiators which are a type of heat exchangers. A system which has a cross-flow type of arrangement is employed in existing plain fins-type radiators which are made up of aluminium and copper alloy. The rate of heat dissipation is improved when a water pump with a powerful fan is supported with it. For a radiator to have a high cooling capacity rate, the addition of fins is one of the methods. By increasing the contact surface this method becomes a key principle. Altering the geometry of a fin can increase the contact surface. Some modifications in the geometry of an existing fin are settled in this chapter for the purpose of improvement of the rate of heat dissipation.

Abdullah H. Alessa et al. [1] made fins rectangular which are horizontal where triangular perforations which are equilateral in nature were fixed firmly for heat transfer enhancement of the natural convection and concluded that the perforated fin which was taken for study amplifies the rate of heat dissipation and also the price of fin material decreases, which ultimately increases the enhancement of transfer of heat for certain values of triangular dimensions. This gives a conclusion that the measure of intensification of heat transfer is in proportion to the thickness of the fin and the material's thermal conductivity [1]. M. S. Sohal et al. [2] from their study also described that the fins displayed an increase in heat transfer coefficient when instead of oval tubes removal of circular tubes was done and when winglets were added to the fins. The addition of winglets caused an increase in heat transfer [2]. In a similar way, Ke-Wei Song et al. [3] observed that the heat transfer was increased when winglets were mounted on the surfaces of the fin and not on the tube bank fin heat exchanger [3]. In the experimental study, J. He et al. [4] experimented with their study in wind tunnel testing of round-tube plain fin heat exchangers where winglet pairs were positioned at an angle of 10° and 30°. Results indicated that the pair of 30° angles of attack produced better performance.

B. Ramdas Pradip et al. [5] reported that thermal systems were used by many industries creating overheating of the components, thereby resulting in the failure of the system, causing damage to the system components. Emitters like fins, ribs, baffles etc. which are powerfully effective in nature were used to solve this problem. "Heat transfer Augmentation" which is also called "Heat transfer Enhancement" or "Intensification" is a term used for the use and development of many techniques which were affecting the energy, material and cost expenditure by increasing the thermal performance of the systems. The transfer of heat by convection thus increases by this technique of Augmentation which reduces the thermal resistance in a heat exchanger. For the increase in heat transfer coefficient, the techniques used for enhancing heat transfer are essential but the cost of an increase in pressure drop should be kept in mind [5].

Yadav J.P et al. [6] also reported that aluminium radiators used by today's modern cars are made by the method of brazing thin aluminium fins which are used to flatten the aluminium tubes. From the inlet, the coolant flows and flows to the outlet through an arrangement of parallel tubes which are mounted [6].

Patel J.R et al. [7] has reviewed applications for modelling the flow of fluid and heat transfer performance characteristics in computational fluid dynamics (CFD) and thus suggested a design constructing a replacement for the conventional radiators of automobiles. Due to the air having the most impact on the overall rate of heat transfer, fins are attached to increase the heat transfer area on the side through which the air is flowing. Some analysis of the properties like the mass flow rate of air, the pitch of the tube and some coolants were done numerically in the computational domain [7].

Jae dong Chung et al. [8] showed and tested the various angle of attack and found that a 30° angle gave better performance by using rectangular winglet pair with a plate heat exchanger which was analysed numerically for examining the combined effects of the louvre fins and vortex generators. [8]. K. M. Kwak et al. [9] investigated the result which found that in order to have better thermal performance the arrangement of fin tubes should be in-line rather than having a staggered arrangement and where the longitudinal vortices are coupled with air-cooled condensers [9]. Yadav Saroj et al. [10] studied a three-dimensional fin system made of aluminium whose air is taken as the working fluid, a thermal analysis by taking different parameters for input keeping in view the different shapes of the fins which have been modelled numerically and where conjugate heat transfer physics has been simulated [10]. Yadav Saroj et al. [11] also studied the triangular-shaped fins considering the varied parametric conditions and its thermal analysis which is numerically analysed by a Finite Element Modelling solver. Accordingly, the weight should be minimised in the system so as to improve the heat transfer rate inside the system which was their main aim. Some parameters which include shape, size and material, orientation, relative arrangement and positioning of the fins, the velocity of working fluids, etc. affects the thermal diffusion in a fin and these effects were analysed on the basis of Nusselt number, coefficient of friction, convective heat transfer coefficient, and coefficient of pressure. In the flow domain, the pressure gradient gets affected due to the extended surfaces. Their study included 3D simulations for the above-stated problem.

The present work is just an attempt to numerically have proof of the effects of different types of fin geometry on the heat transfer rate of radiators on a computational domain. Due to smaller rates of heat dissipation in the automotive radiators, there has been a problem coping with the demand for smaller spaces of a hood kind creating a problem of overheating of the engine, and ultimately leading to the breakdown of engine parts and creating stress due to the significant wear occur in metal parts owing to the evaporation of lubricating oil. To decrease or minimize this problem in the engine which occurs due to the heat generation, the automotive radiators were redesigned in a compact form which still maintains the performance of heat transfer at high levels. One of the reliable solutions is redesigning the fin for the current problem and the current work which is specifically dealing with several fin designs given under similar parameters of study.

This chapter focuses on the study of the different elements like the shape of the fins, space between the fins, tube size, the material used in fins, air inlet temperature through COMSOL MULTIPHYSICS 5.5 and also the shape of the radiator core, the study of the working fluid whose direction of flow is analysed along with the radiators' frontal area taken from several other research papers and eventually forming an optimized geometry through the various parameters.

The objective of this study is to investigate the different kinds of geometric configurations and associated additional fin attachments to increase the efficiency of the cooling system which further prevents detonation or pinking and pre-ignition etc. that reduces the energy efficiency of the vehicle and also causes serious mechanical problems to the vehicle. The need for this work is that only 15% of the combustion energy is used for the vehicle's movement on average while the large chunk of it is wasted as heat and with time as the engine further gets heated it reduces the engine efficiency drastically, hence, an efficient cooling system is necessary for automobiles which also reduces pollution by increasing energy efficiency.

This chapter deals with the four designs and configurations of radiator fins in order to have a comparative data analysis on the basis of effectiveness, efficiency and temperature variation in different fin geometries like simple rectangular radiator fin, adjoint additional squared fin, circular fin and radiator with changed tube diameter respectively. The whole analysis was conducted computationally using COMSOL Multiphysics 5.5 software. The result showed that the efficiency increased with the addition of square and circular fins with a comparatively higher rise in the case of additional circular fins that is 59.99%, 60.4%, 63.78%, and 70%, respectively, and we also noted that the temperature variation alongside the length of the fin side for inlet of air achieves higher peaks and hence gets heated more affecting the effectiveness.

7.2 FINS

In the heat transfer study, the extended surfaces from an object commonly called fins, help in expanding the rate of heat transfer to the environment or from the environment thereby assisting in increase of the rate of convection. The total amount of heat which is transferred is determined by the conduction, convection or radiation of an object. The convective heat transfer coefficient will increase if the gradient of the temperature profile between the environment and the object is maximized. There is a main solution to certain problems in heat transfer which can be solved if the surface area is increased while adding a fin to an object.

Applications of fins generally include heat transfer purposes for equipment such as heating instruments or cooling instruments. So, calculating the rate of heat transfer and the temperature of the fin by analysing it has become an important step. Transient thermal analysis gives the correct value of temperature which is varying with time. For the analysis of fins, selecting the correct material is an important factor. Various types of materials like aluminium, copper, carbon steel, brass, etc. are used. Engineering fins are also used as heat transfer fins for the regulation of temperature profiles in heat sinks or fin radiators.

7.2.1 Types of Fins

Fins which are classified as longitudinal, radial and pin fin are used for numerous applications. The profiles for longitudinal fins consist of concave parabolic, rectangular profile and trapezoidal. The profiles comprising radial fin consist of triangular and rectangular profiles and finally, the profile comprising pin fin consists of tapered type, cylindrical type and concave parabolic profiles.

7.2.2 TYPES OF FINS USED IN RADIATORS

Radiators may be made with plate fins and serpentine fins. In plate fins, tubes are inserted through stacks of relatively flat fins that have tube holes in them while in serpentine fins rows of tubes are stacked with layers of corrugated fins. Whether radiator fins are plated or serpentine types, they may be louvred or non-louvred [12].

7.2.3 CONDUCTION

In the branch of physics, heat is transferred other than by work or by the transfer of matter and can be defined as the quantity of energy. Between two bodies, the transfer of energy as heat is a spontaneous process that occurs when the bodies have a physical connection and differ in temperature. The transfer can occur by conduction, radiation and convection. In the form of electromagnetic waves, heat transfer through radiation takes place which is mainly in the infrared region.

Conduction is called heat transfer by way of continuous random movement of atoms and molecules precisely, molecular agitation due to thermal energy. It happens without any motion of the material as fully within a material. By diffusion of microscopic particles and the materials collide, the heat is transferred as a whole and also by collision between particles which are called quasi-particles inside the body which is due to temperature gradient and diffusion of microscopic particles. The substance towards the colder end will receive energy which will be transferred down if one end of the substance is at a higher temperature, this is because of the reason that the hot particles which have a higher speed than the colder particles have a collision effect with the colder particles which has a slower speed giving a net transfer of energy to the colder particles as they are slower in speed. In all four forms of matter including solids, liquids, gases and plasmas, conduction plays a significant means in heat transfer. Fluids and gases have less conductivity when compared to solids, (gases are weaker in conductivity), and the reason behind this is that the atoms in a molecule of gas have larger distances between them which leads to fewer collisions between the atoms, ultimately leading to a decrease in conduction. If temperature is increased the conductivity of gases is increased.

7.2.4 FOURIER'S LAW OF HEAT CONDUCTION

The statement of Fourier's law, which is the law of conduction of heat, gives the rate of heat transfer with time through a material which is in proportion to the temperature and area through which the heat flows; temperature has a negative gradient and the area makes 90° to that gradient. This law can be described in two ways. Firstly, the determination of the amount of energy which flows in or out of a whole is called its integral form, and the value of fluxes or flow rates is called the differential form. This law gives us a relation between the local heat flux density, $\bar{q}$ which is equal to the product of thermal conductivity, k and the negative local temperature gradient, (dT/dX). The amount of energy flowing through a unit area is the heat flux density which is defined in the case of per unit time. In mathematical terms, this can also be shown as - $\vec{q} = -k\,(dT/dx)$

Suppose if the material's total surface is denoted by S, then by integrating the differential form over S, the relation which is obtained is:

$$dQ/dt = -k \int_{s} \vec{\nabla} . T . dA. \qquad (7.1)$$

Where, dQ/dt stands for the rate of flow of heat.

7.2.5 THERMAL CONDUCTIVITY OR CONDUCTIVITY CONSTANT OR CONDUCTION COEFFICIENT, k

The measure of a material at what amount it is significantly able to conduct heat to another material where temperature, density, its phase and bonding between molecules are dependent on it is defined as the thermal conductivity of that material. It describes the ease of conduction of the system. Denoted by k, this unit is assessed by Fourier's law for the conduction of heat. It is stated as the quantity of heat (Q), which makes normal to the surface of area (A) in terms of direction, which is through a thickness (L), and transmitted with a time (t) which happens because of a temperature difference (ΔT). Materials having higher thermal conductivity, have a high heat transfer rate in comparison to the lower thermal conductivity materials. The SI unit of thermal conductivity is watts per meter kelvin (W/ m.k).

7.2.6 CONVECTIVE HEAT TRANSFER

The transfer of heat between two bodies by the movement of fluids or gases is termed convective heat transfer or convection. Convection happens usually when heat transfer occurs in liquids and gases. Convection is a definite method of heat transfer which comprises two processes, i.e., conduction which is not known, also called heat diffusion and heat transfer mechanism through a fluid by bulk fluid motion which is termed advection.

Newton's law of cooling describes the convection-cooling process. Whereas, Newton's law states that breeze creates some effects and under that, the body at which rate it loses the heat is in proportion to the body temperature and its surroundings making temperature differences between them. The constant of proportionality is actually the heat transfer coefficient. When there is a difference in temperatures between the object and surrounding and its coefficient of heat transfer is relatively independent then this law is applied.

When there is a smaller change in temperature Newton's law gives an approximate reality. The dependence of the constant of proportionality of heat transfer on temperature is direct and is in standard form.

The convective heat transfer has the relationship of:

$$Q = hA(T - T_f) \qquad (7.2)$$

Here Q refers to the heat transfer rate, the heat transfer coefficient is denoted by h, the area of the body is denoted by A, the surface temperature of the object is denoted by T and the temperature of the fluid is denoted by T_f.

Convective heat transfer coefficient depends upon the properties which are physical in nature. For fluidic flow situations, the common values of h have been tabulated and also measured.

7.2.7 CONVECTIVE HEAT TRANSFER COEFFICIENT

Flow properties which include velocity, temperature and viscosity along with the type of media where the fluid is flowing are the two properties where convective heat transfer coefficients, which are denoted by h_c, are dependent.

7.2.8 CONVECTIVE HEAT TRANSFER COEFFICIENT FOR AIR

For air flow, the convective heat transfer coefficient has the relation of:

$$h_c = 10.45 - v + 10\, v^{1/2} \qquad (7.3)$$

Where, The heat transfer coefficient is denoted by h_c whose unit is in kCal/m^{2}h°C and The relative speed between the object surface and the air is denoted by v whose unit is in m/s.

We know, 1 kcal/m^{2}h°C equals 1.16 W/m^2°C
Then Equation (7.3) can be evaluated to give

$$h_{cW} = 12.12 - 1.16\, v + 11.6\, v^{1/2} \qquad (7.4)$$

Here, the heat transfer coefficient is denoted by h_{cW} whose unit is W/m^2°C. The above equation to be noted here can be used for velocities ranging between 2 to 20 m/s. and it is an empirical equation. [13]

7.2.9 THERMAL RADIATION

Thermal radiation is another property of heat transfer of matter which is generated as energy in the form of electromagnetic radiation when there is a random motion in the particles like atoms, molecules, etc. of matter. If the temperature becomes higher than absolute zero, all these particles emit thermal radiation. This phenomenon can be understood by some examples which include that the animals which can emit infrared radiation are detectable with a thermal imaging camera and also the radiation caused by the background of the cosmic microwave. In thermodynamic equilibrium, when an object having the physical features of a black body emits a certain kind of radiation then that kind of radiation is termed as the black body radiation. Depending only on the temperature of an object, the continuous frequency spectrum produced by black body radiation is expressed by Planck's law. Wien's displacement and Stefan-Boltzmann give two laws related to radiation. The former governs the frequency of the radiation which is emitted, and the latter gives the radiant flux per unit solid angle of the radiation.

The rate of energy transfer from one surface to the second surface for a black body is given by the following equation:

$$Q_{1 \to 2} = A_1 E h_1 F_{1 \to 2} \tag{7.5}$$

7.2.10 FIN EFFICIENCY

The efficiency of fins is determined by the heat transfer rate which is actual and which takes place through the fin making a ratio with the maximum possible rate of heat transfer that can occur through the fin. By this, it explains that the complete fin is in its base temperature or room temperature. The complete fin which has its material having infinite thermal conductivity, will be in its base temperature only. Mathematically, the below equation gives the efficiency of fin.

$$\text{Efficiency, } \eta = Q_{actual}fin / Q_{ideal}fin \tag{7.6}$$

Where,

$$Q_{actual}fin = h \, A_{fin} (T_{av} - T_s) \tag{7.7}$$

$$Q_{ideal}fin = h \, A_{fin} (T_{max} - T_s) \tag{7.8}$$

7.2.11 FIN EFFECTIVENESS

The actual rate of heat transfer with the fin making a ratio with the rate of heat transfer without the fin gives the effectiveness of the fin. Mathematically, the below equation gives the effectiveness of fin:

$$\text{Effectiveness, } \epsilon = Q_{actual}fin / Q_{without}fin \tag{7.9}$$

Where,

$$Q_{actual}fin = h \, A_{fin} (T_{av} - T_s) \tag{7.10}$$

$$Q_{without}fin = h \, A_{withoutfin} (T_{max} - T_s) \tag{7.11}$$

The three conditions for fin effectiveness include:

i. When $\varepsilon = 1$: the heat transfer doesn't get affected by the fin.
ii. When $\varepsilon < 1$: the fin will have an insulating nature when the material's thermal conductivity (k) is low.
iii. When $\varepsilon > 1$: the heat transfer rate increases.

Note:

1. If the values corresponding to the convection heat transfer coefficient (h) have lower values, then fins are generally used i.e., when heat transfer is by natural convection and gas or air is the medium.
2. The material of the fin should have a higher conductivity rate.
3. Aluminum is a great choice because it is inexpensive and has lower weight and also aluminium is resistant to corrosion.
4. It is better when the fin's Lateral surface area P/A_c is high.
5. If fins have triangular profiles or parabolic profiles which contain material of lesser amount and require minimum weight then fins are more efficient in nature.

Fins used in practice have efficiencies above 90%.

7.3 RADIATORS

A radiator is a kind of heat exchanger transferring thermal energy between a source and a working fluid in the process of heating as well as cooling. A radiator consists of a bigger cooling surface for efficient cooling through water which contains enormous amounts of air. The radiator is mainly used in automobile industries. In automobiles, the internal combustion engines get heated up so to cool it, radiators are used. They are also used in different sectors like railways, motorcycles, aircraft having piston engines, locomotives, plants which make stationary objects and many more places. The direction of the flow of water through radiators gives the general classification for types of radiators. When water vertically flows from top to bottom then that kind of radiator is known as the down flow type. Whereas, when water from a tank on one side flows horizontally to a tank on the other side, on this basis the type of radiator is known as the cross flow type. Copper and brass have high thermal conductivity, therefore these materials are used to make radiators. By the process of soldering the sections of the radiators are joined [14].

7.3.1 Types of Radiators

The two basic types of radiators include:

1. Tubular type core.
2. Cellular type core.

7.3.1.1 Tubular Type Core

In this type of radiators, there are a series of tubes where the fins are attached around them which improves the rate of heat transfer connecting upper and lower tanks through which the water flows. The mechanism of this type of radiator is that the air flows through outside of the tubes and also between the fins thereby the water which is flowing inside the tubes absorbs the heat. In this kind of radiator, the tube will lose its effect of cooling if one of the tubes gets obstructed since the water flows through all the tubes.

7.3.1.2 Cellular Type Core

In this kind of radiator, a very large number of individual air cells are filled in core material which is surrounded by water and the flow of water is in the spaces between the tubes. The air passes through the tubes. The cellular-type core radiator appears like a honeycomb when the hexagonal form of cells is in front, so it's also called a honeycomb radiator. In this kind of radiator, the cooling effect will be lost but for a very small part of the overall cooling surface of the radiator if there is any clogging in any passage of the tube.

7.3.2 CONSTRUCTION OF RADIATOR

A radiator has an upper tank which is connected by a hose pipe to the outlet through which water will flow from the engine jacket, a lower tank which is connected through the pump where water flows to the inlet of the jacket, and a core. The core is placed between the upper tank and lower tank of a radiator. The core of the radiator cools the water which is a radiating element in nature.

7.3.2.1 Principle in which a Radiator Works

In a radiator, the working principle is not a complex process. In the radiator, the arrangement of tubes is parallel. Through these tubes the coolant flows. The flow is from the inlet to the outlet. Through the inlet port of the radiator, behind the radiator, the hot water enters and a fan is attached there for the purpose of cooling down the heat produced as hot water is flowing inside the tubes. The water cools down from the air blown by the fan and thereby the water temperature decreases at the outlet and then it will be sent back to the engine.

The mechanism of a radiator is very simple. Currently, modern cars have radiators whose material is made up of aluminium. On each side of the Radiators, there are tanks, and a cooler transmission is fitted inside the tank. Here an aluminium mesh type of radiator is used. This device has two ports, i.e., inlet port and outlet port. Fins made of aluminium material are attached to the tubes which are inside the radiator and are parallel arranged with the radiator.

For the application of flowing air, fins made of aluminium which are attached to the tubes are known as tabulators.

The aluminium coating receives heat from the air passing from the fan when the hot coolant comes from the engines and fills the tubes by which the aluminium coating gets cooled. In the spaces where the flowing fluid touches the tube, the part touching the tube will be directly cooled which will again go back to the engine but before that, it will be sent out to the cooler.

7.4 COMSOL

COMSOL Multiphysics is a software which is used to do simulation along with designing processes or any devices in fields of manufacturing, research and engineering applications. This software's user interface is conventional which is physics-based and used for solving partial differential equations (PDEs) and does analysis of finite elements and acts as a solver providing an integrated development environment of multiphysics provided with a unified workflow for numerous applications in the field of different engineering or physics-based applications.

Some classical problems can also be addressed with some application modules but generally, PDEs are solved weakly in this multiphysics package. For COMSOL, many modules are present which are classified according to the particular application areas. The categories are divided into fields in mechanical, fluids, acoustics, electrical, chemical, interfacing and multipurpose [15].

7.4.1 INTRODUCTION TO FINITE ELEMENT METHOD

In physics, partial differential equations (PDEs) are considered in terms of time-dependent and problems of space. In most of the designs for making geometries or for any different problem there is no option for solving PDEs systematically. So, the solution to these kinds of problems is that we can approximate the equations which will be done with different discrete counterparts called discretization. The first step of this method is to make them able to be solved or implemented by numerical evaluation where the PDEs will be approximated by methods of discrete parts by some numerical model equations. An approximation to the real solution of PDEs is given by these model equations. Thus, the finite element method (FEM) is used for solving such problems where difficulties are faced in finding solutions to some problems by computing such approximations.

For example, a variable dependent on a PDE has a function u (i.e., temperature, electric potential, pressure, etc.). By basic function using a function which is a combination of linearity a function u_h gives an approximation to the function u which is in accordance with the following expressions:

$$u \approx u_h \tag{7.12}$$

$$u_h = \sum u_i \psi_i \tag{7.13}$$

Here, ψ_i denotes the functions that are basic in nature and u_i denotes the coefficients of the functions that give an approximation to u with u_h. Finite element methods have their own benefits, one of them includes that the finite element method has its own choice of selecting the discretization. In elements that can use basic functions or can discretize space, both of them can be selected. The second benefit of this method is that it provides a relationship theory between numerical formulation and weak formulation of the problems of PDE which is better understood. For instance, when the model equations are solved numerically on a digital platform the theory gives the estimation of errors or bounds for the errors thus becoming a well-developed theory [16].

7.5 METHODOLOGY

The solution of Computational Fluid Dynamics calculations, usually comprises three main steps: the first one is *Pre-Processing,* second step is called *solution phase,* and finally the *Post-Processing* part. The pre-processing step comprises the step where discretization is created and solved into a finite element by giving certain boundary conditions and developing equations for the element. The solution phase or the solver execution phase is the step where equations used to obtain nodal results with certain boundary conditions are solved and starts the solver which runs until the entire convergence is reached finally, the post-processing part includes the results which are examined when the solver is terminated.

Step 1. A 3D model with fins inside is drawn in the Radiator section by giving proper dimensions for different models which is done in COMSOL MULTIPHYSICS 5.5.

Step 2. The created fins in COMSOL software are used for further simulations.

Step 3. The data obtained is used for the analysis and further calculation of efficiency, effectiveness and other parameters.

7.6 MODELS WITH SPECIFICATIONS

7.6.1 FIN MATERIAL

Heat transfer characteristics play a major role in determining the materials which will be used in fins. The materials generally used are a mixture which is an Alloy of aluminium which has a number A204 generally having thermal conductivity ranging from 110–150 W/mK, and another alloy of aluminium which has a number of 6061 also has a higher thermal conductivity. Alloy made from cast iron and copper is also a material used in making fins. Since heat exchangers play an important role in their value in the environment, the need to require a range of construction materials rises. The standard and most used fin material is aluminium and the other important fin material is copper.

Natural Aluminium: The most standard fin material is natural aluminium. It's generally used in electrical coils and aluminium alloys are used in making fins in heat exchangers. To fight corrosion, these fins are coated with standard finish corrosion protective coating. And also gives a solution for all conditions being in-expensive. The fin size is also an important matter in the production. The demand for thinner fin stock materials is higher because of the decrease in weight of the fins, and for this reason, the need to improve the mechanical properties making it resistant to corrosion and having good thermal conductivity has widely grown.

Copper: In application areas where there is no alternative in replacement of the heat exchangers, copper becomes the ideal choice for demanding environments, providing the utmost resistance to corrosion of the different kinds of equipment used in coils improving the life of the whole system. Copper has thermal conductivity, corrosion resistance and also superior strength, as compared with aluminium. Heat exchanger fins made of copper material are essential for becoming environment friendly where there is a need for improved thermal performance. Copper fins are suited to more abrasive environments because it is difficult for fins made of copper to wear out and fail, which doesn't happen with a coated fin [17].

Aluminium Properties used in the models:

$$Density\ of\ aluminium\,(\rho) = 2700\ kg/m^{3},$$
$$Specific\ heat\,(C_p) = 0.\ 871\ kJ/kg.K,$$
$$Thermal\ conductivity\ of\ Al\ (k) = 202.4\ W/m.K.$$

7.6.2 Specifications of Models

The four different configurations of fins used in the study are illustrated in Figure 7.1 whose configuration details are given below:

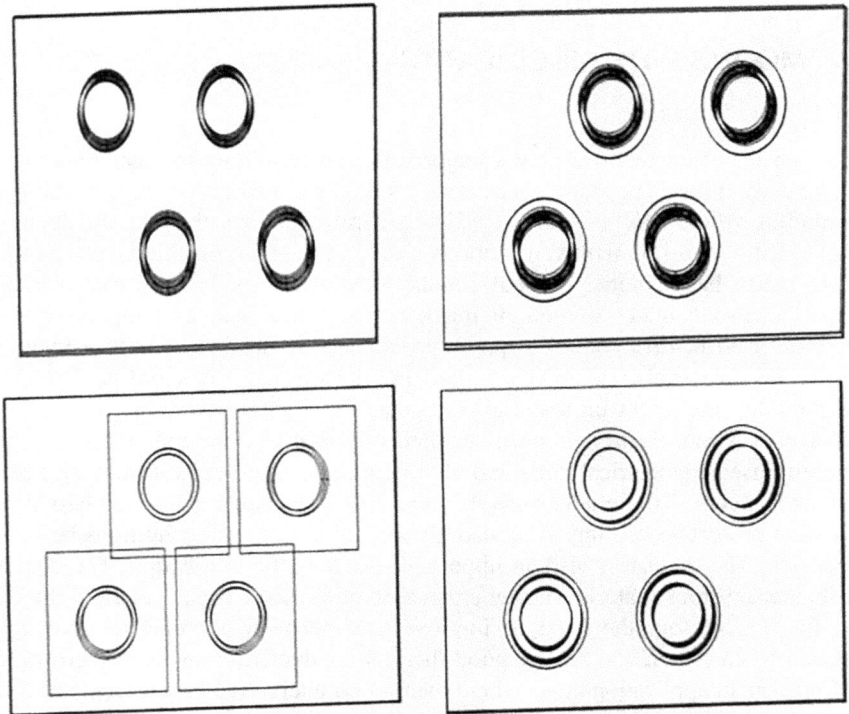

FIGURE 7.1 Different configurations of fin types.

7.6.3   MODEL 1.0

Radiator Model consisting of simple rectangular radiator fin. [18]

Type of Radiator	= staggered type
Number of tube rows (N)	= 25
Number of tube columns	= 2
Radiator material	= Aluminium
outside diameter of tube (d_o)	= 6.35 mm
inside diameter of tube (d_i)	=5.35 mm
Thickness of tube (T_t)	= 0.5 mm
tube Pitch longitudinal (S_L)	=12 mm
tube pitch transverse (S_T)	=13.87 mm
tube pitch diagonal (S_D)	= 13.87 mm
Fin thickness (F_t)	= 0.3 mm
Fin spacing (F_S)	= 1.60 mm
Fin pitch (F_P)	= 1.90 mm
Length of radiator (L)	= 300 mm
Depth of radiator (D)	= 28 mm

7.6.4   MODEL 2.0

Radiator model consisting of adjoint additional circular fins.

This model has the same dimensions as model 1.0. Only, it has an additional circular fin which has a cylindrical fin diameter = 9.75 mm.

7.6.5   MODEL 3.0

Radiator model consisting of the adjoint additional squared fin.

This model has the same dimensions as model 1.0. Only, it has an additional squared fin whose length if the squared fin = 12.7 mm.

7.6.6   MODEL 4.0

Radiator model comprising of simple adjoint additional circular fins radiator fin with a changed tube diameter. This model has the same dimensions as model 2.0. Only, it has additional circular fins like model 2.0 but whose tube outside diameter is changed to 7.54 mm.

7.6.7   The Boundary Conditions and Governing Equations Include

The equations solved are time-dependent flow equations and flow is considered laminar. The fluid flow condition and heat transfer have governing equations consisting of continuity equations (here viscous dissipation is neglected) which are incompressible in nature; the Navier– stokes equations in X, Y and Z direction momentum equation, energy equations and with the equation of state.

The X, Y, Z Momentum Equations consists the following equations:

$$
\left.
\begin{aligned}
\nabla(\rho u \forall) &= -\frac{\partial p}{\partial x} + \frac{\partial \tau_{xx}}{\partial x} + \frac{\partial \tau_{yx}}{\partial y} + \frac{\partial \tau_{zx}}{\partial z} + \beta_x \\
\nabla(\rho v \forall) &= -\frac{\partial p}{\partial y} + \frac{\partial \tau_{xy}}{\partial x} + \frac{\partial \tau_{yy}}{\partial y} + \frac{\partial \tau_{zy}}{\partial z} + \beta_y \\
\nabla(\rho w \forall) &= -\frac{\partial p}{\partial z} + \frac{\partial \tau_{xz}}{\partial x} + \frac{\partial \tau_{yz}}{\partial y} + \frac{\partial \tau_{zz}}{\partial z} + \beta_z
\end{aligned}
\right]
\tag{7.14}
$$

The Energy Equation:

$$
\nabla(\rho u \forall) = -p \nabla \, \forall + \nabla(k \nabla T) + \phi + S_h]
\tag{7.15}
$$

And finally, the Equation of State:

$$
p = \rho R T]
\tag{7.16}
$$

The computational domain of the models for the study consists of the dimensions which are having length, height and width equal to l = 41.8 mm, h = 28 mm and w = 4.2 mm. Figure 7.2 shows the computational domain of the reference model. Here, the base area of the cooling fin is the same for all the models. The inlet velocity Vi and the inlet temperature Ti have values: Inlet velocity of air V varies (V1 = 5.1920 m/s, 19.145 m/s, 15.576 m/s, 20.768 m/s, 25.968 m/s) and by taking uniform assumption through the whole boundary of inlet temperature which is equivalent to Reynolds Number of the range 1000 to 5000. The inlet temperature of

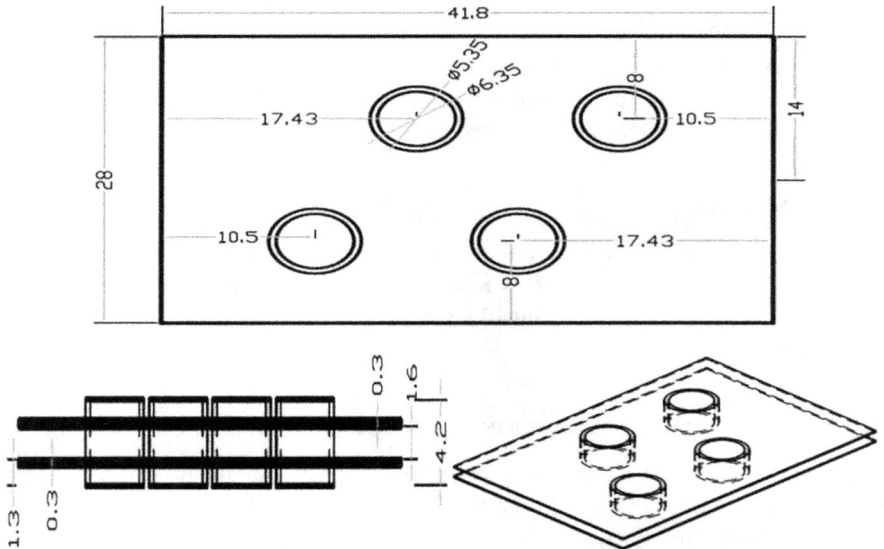

FIGURE 7.2 Schematic of the reference model's computational domain.

T_{in} has a value of 303 K (Pr = 0.7). The inlet velocity of water having Smooth water of the tube V (has a value of V1 = 5.1920 m/s) and the inlet temperature T_{in} are assumed uniform through the entire boundary. Hot incompressible fluid has a temperature of 363 K of water in all cases. The first model's shape has circular, oval tubes and the fin spacing on the model is 1.6 mm.

Figure 7.2 shows the computational domain, which shows the computational analysis by showing the schematic model. The model has the basic dimensions of a 3D model. The domain consists of fins and tubes consisting of water and air. The domain which has fluid consists of the air flow volume and for a double fin and tube assembly, a coolant flow volume is created because of the geometric similarity between the fins and tube rows which ultimately helps us in making the computational domain have an arrangement of having four tubes and adjoining fin with it and finally analysis of design is carried out. The solution to the problem is a conjugate heat transfer which makes it necessary to know the thickness of the tube and fin.

7.6.8 INLET CONDITIONS OF FINS

The inlet temperature of air is taken as 303.15 K, the viscosity of air (μ) has a value of 18.63×10^{-6} Ns/m^2, thermal diffusivity (α) is equal to 22.861×10^{-6} m^2/s, density of air (ρ) has a value of 1.165 kg/m^3, Prandtl number equals to 0.701, the thermal conductivity of air (k) is taken as 0.02675 W/m.K and the value of specific heat (Cp) of the material is 1.005 kJ/kg.K.

The inlet temperature of the water is taken as 363.15 K, the viscosity of water (μ) has a value of 0.315×10^{-3} Ns/m^2, thermal diffusivity (α) is equal to 0.16585×10^{-6} m^2/s, density of water (ρ) has a value of 967.5 kg/m^3, specific heat (Cp) of the material is 4.2055 kJ/kg.K, Prandtl number equals to 1.98 and thermal conductivity of air (k) is taken as 0.67455 W/m.K.

7.6.9 FLOW CONDITIONS – PERIODIC

The flow condition which is periodic in nature splits its selection into two groups, i.e., a source group and a destination group. Through one of the destination boundaries, the fluid leaves the domain and enters through the corresponding source boundary. This situation shows that the geometry is a periodic part of a larger geometry. The velocity vector will automatically be transformed if the boundaries are not parallel to each other [19].

Velocity at inlet = velocity of outlet
Pressure at inlet = pressure at outlet
Air inlet velocity =19.145 m/s
Water inlet velocity = 5.1920 m/s

In the COMSOL model, the heat transfer and fluid flow condition has the spacing of the fin considered to be air domain, and the tube has a domain of hot water, the thickness of the fins along with the thickness of the tube are having the same mesh

TABLE 7.1
Element Counts in Tube Fin Assemblies for the Models

Statistics	Model 1.0	Model 2.0	Model 3.0	Model 4.0
Tetrahedral	253625	431506	438379	491476
Pyramids	4640	5568	5568	5992
Prisms	56786	82028	82028	87752
Triangles	72958	112228	120502	116060
Number of elements	315141	519102	525975	585220
Average element quality	0.6175	0.6367	0.6304	0.6415
Mesh volume	2515mm^3	2515mm^3	2515mm^3	2515mm^3

density. For perfect accuracy, the grid independence study is solved in order to decrease the requirement of the number of elements. The finalized element counts and the related aspects are listed below in Table 7.1.

7.7 RESULTS AND DISCUSSIONS

7.7.1 Results Obtained from Model 1.0

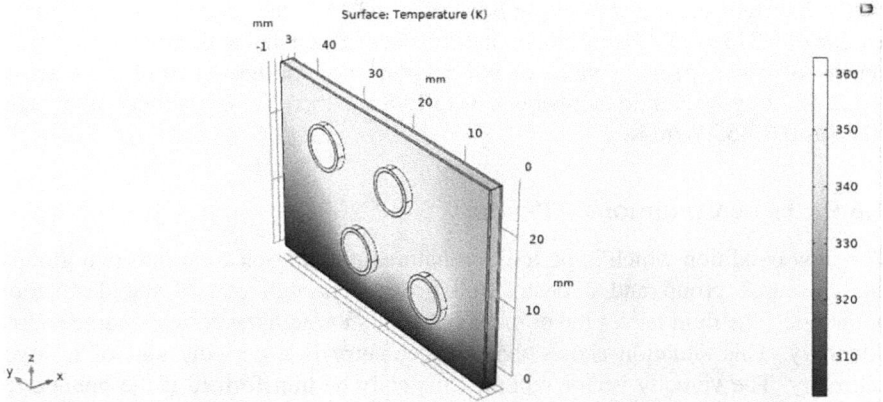

FIGURE 7.3 Surface temperature distribution at the fin surface.

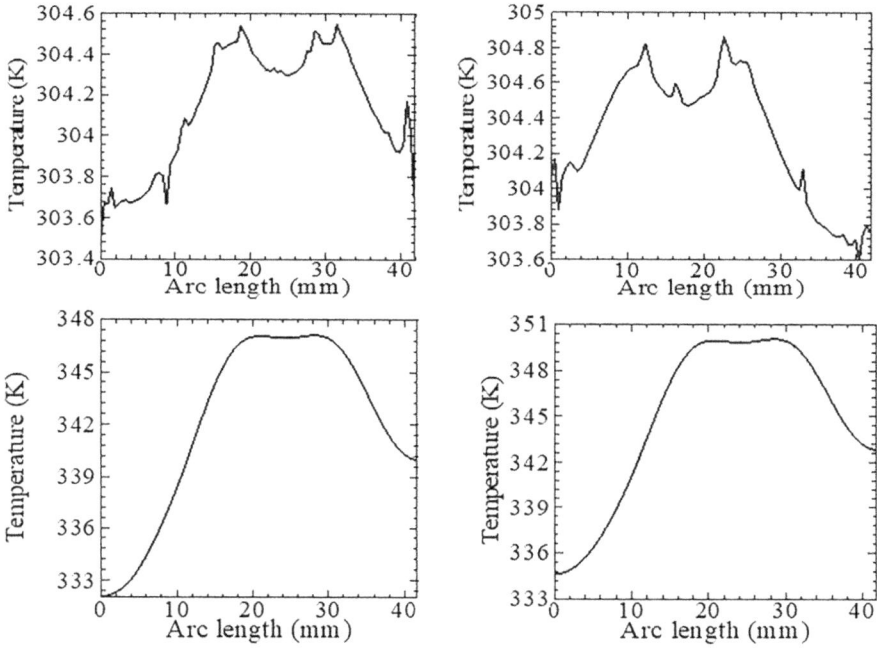

FIGURE 7.4 Graph representation of temperature variation from air inlet and air outlet for the four edges of model 1.0.

7.7.2 RESULTS OBTAINED FROM MODEL 2.0

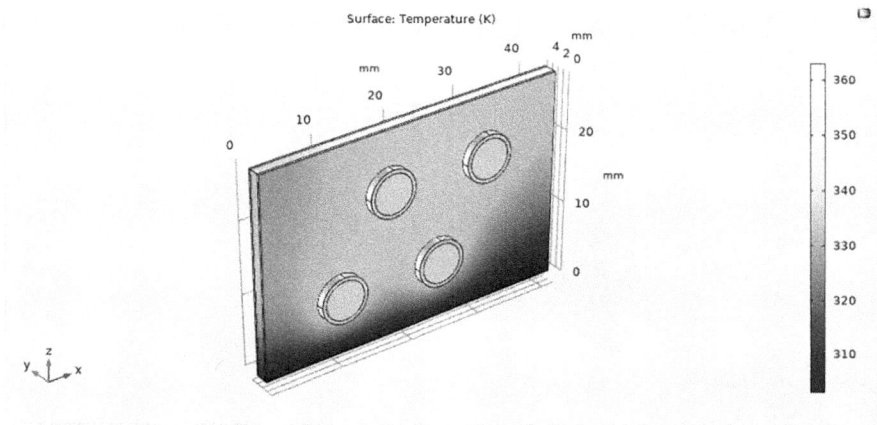

FIGURE 7.5 Surface Temperature distribution at the fin surface.

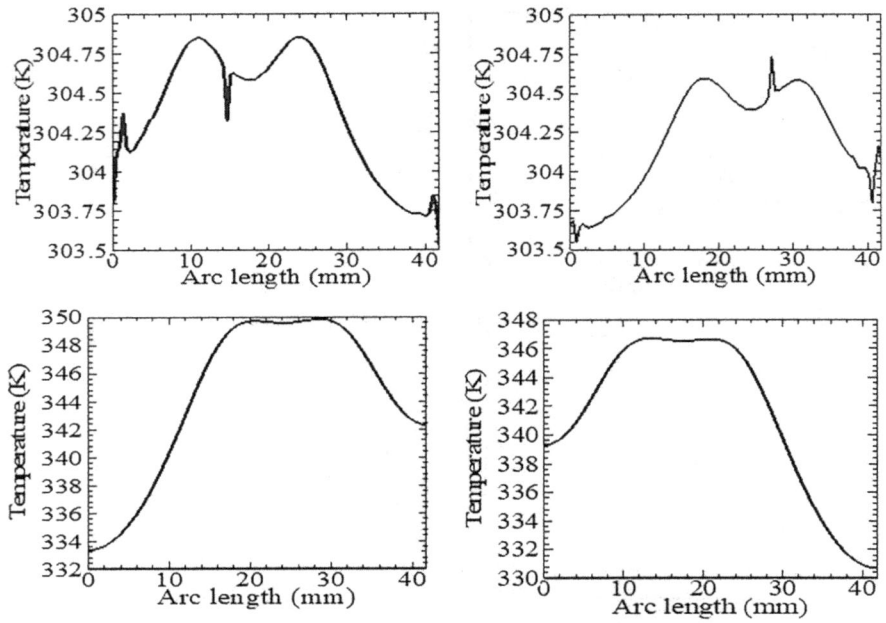

FIGURE 7.6 Graph representation temperature variation from air inlet and air outlet for the four edges of model 2.0.

7.7.3 RESULTS OBTAINED FROM MODEL 3.0

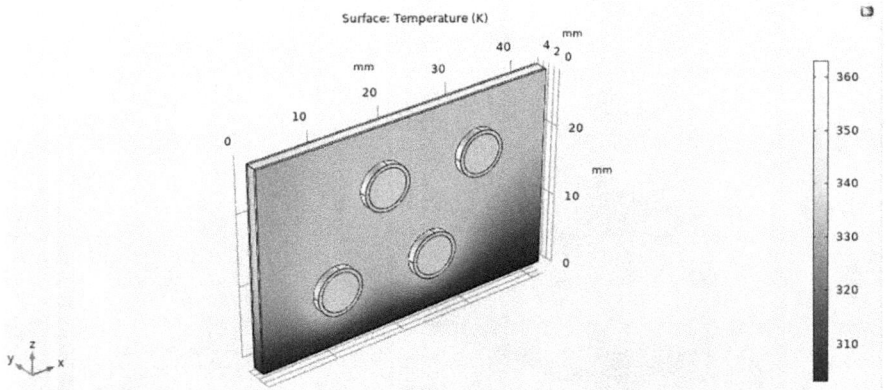

FIGURE 7.7 Surface Temperature distribution at the fin surface.

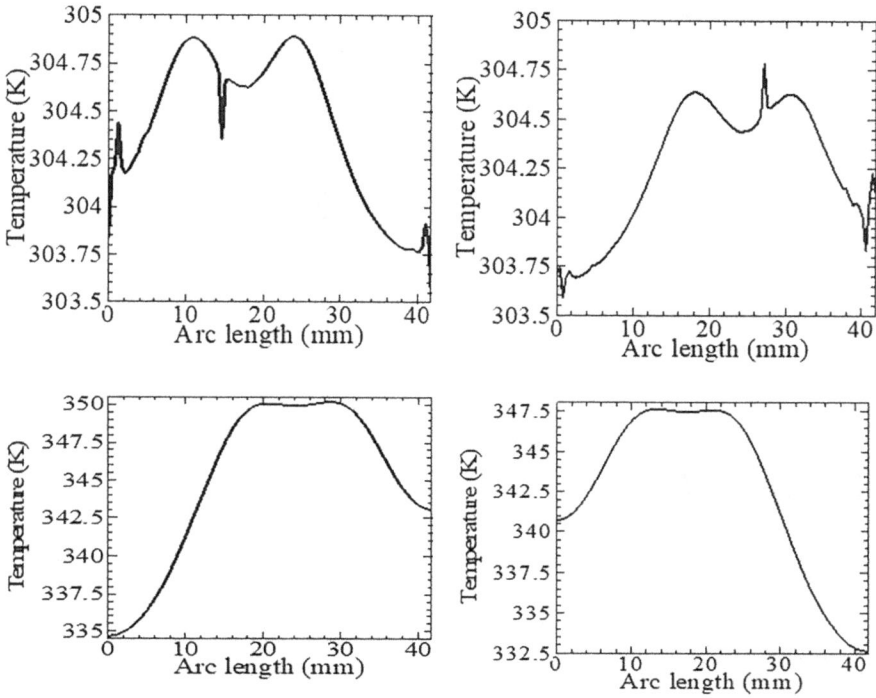

FIGURE 7.8   Graph representation temperature variation from air inlet and air outlet for the four edges of model 3.0.

7.7.4   RESULTS OBTAINED FROM MODEL 4.0

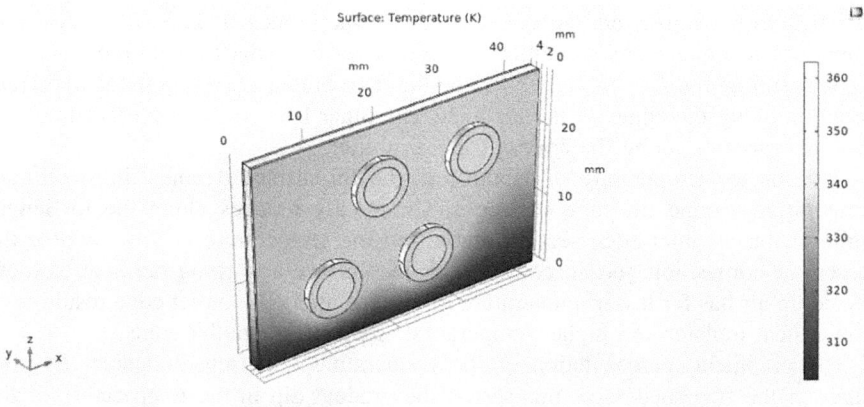

FIGURE 7.9   Surface Temperature distribution at the fin surface.

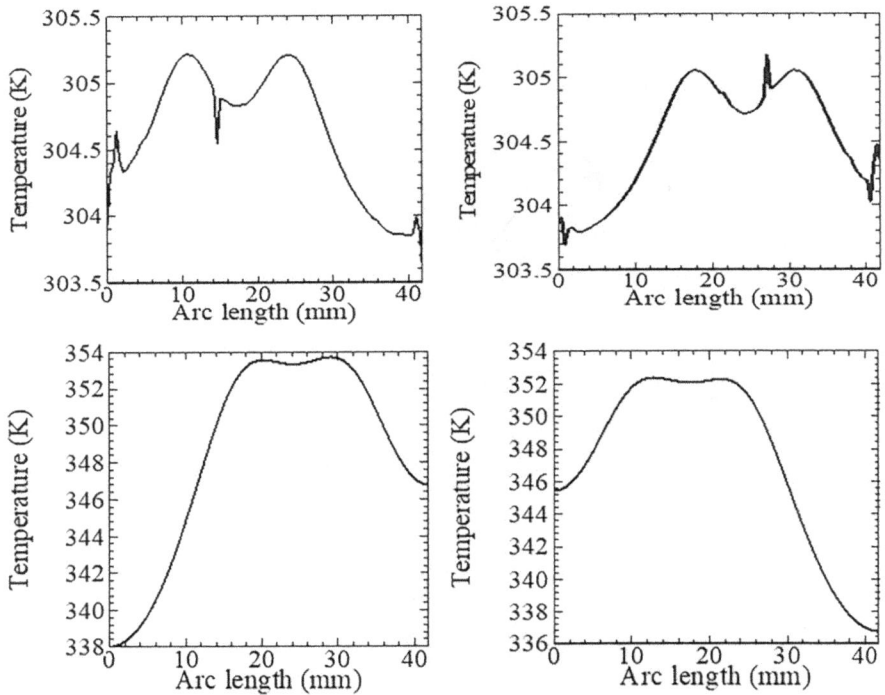

FIGURE 7.10 Graph representation temperature variation from air inlet and air outlet for the four edges of model 4.0.

7.7.5 DISCUSSIONS

Figures 7.3, 7.5, 7.7 and 7.9 denote the surface temperature distribution and Figures 7.4, 7.6, 7.8 and 7.10 denote the graphical representation of temperature variation from air inlet and air outlet edges for the models 1, 2, 3 and 4. Air inlet comprises two edges and air outlet also comprises two edges for all the simulated models. In Figures 7.4, 7.6, 7.8, 7.10, the first two graphs represent the temperature variation along the edges of the air inlet. The other two graphs represent the temperature variation along the edges of the air outlet.

The surface temperature distribution at the fin surface denotes the maximum temperature around the tube surface and gradually reduces along the fin length while at the air inlet edge section of the fin, the temperature is lower where the forced air convection gets in contact with the surface absorbing the heat as these heated up air has far lesser temperature difference along the outlet edge resulting in slower heat transfer and higher temperature along the air outlet edge.

The graphical representation of the temperature variation alongside the inlet edges of the two successive fins shows the gradual dip in the temperature of the coolant as it moves down, with higher peaks at around the edges near to the radiator tubes, the variation for model 1.0 is effectively shown in Figure 7.4, while for outlet there is a similar dip on the coolant temperature as it moves down the fins, the graphs for air outlet also shows the near uniform high temperature along the outlet edge as has already been shown in Figure 7.3, for the surface temperature distribution at the fin surface.

Figure 7.6 shows that the addition of mid fins in the radiators significantly reduces the drastic temperature variation alongside edges and leads to near-gradual variation while the outlet edge shows an excessive dip of temperature along one of the outlet edge corners. This shows the use of mid-fins does smoothen the temperature variation along the surface as well as enhances the heat transfer. It was also noticed by analysing the graphs of the average temperature of the fin with air inlet temperature that the average temperature of the fin increases linearly with an increase in air inlet temperature; here air inlet temperature is basically the surrounding temperature. Increasing the surrounding temperature very near the fin increases the rate of convection between the fin and the surrounding temperature, which results in an increase in the temperature of the fin.

7.7.6 EFFECTIVENESS AND EFFICIENCY

Surrounding temperature (T_s) = 303.15 K

Maximum temperature/Source temperature (T_{max}) = 363.15 K

To find the heat convection coefficient of air,

$h = 12.12 - 1.16V + 11.6\ V^{1/2}$ [Adapted from https://www.engineeringtoolbox. com/convective-heat-transfer-d_430].

$$V = 19.145 \text{ m/s (in all cases)}$$
$$h = 12.12 - 1.16(19.145) + 11.6(19.145)^{1/2}$$
$$= 40.6675\ W/(m^2 K)$$

$$Q_{actual}fin = h\ A_{fin}(T_{av} - T_s)$$

$$Q_{without}fin = h\ A_{withoutfin}(T_{max} - T_s)$$

$$Q_{ideal}fin = h\ A_{fin}(T_{max} - T_s)$$

$$\text{Effectiveness, } \epsilon = Q_{actual}fin/Q_{without}fin$$

$$\text{Efficiency, } \eta = Q_{actual}fin/Q_{ideal}fin$$

The efficiency and effectiveness of all the simulated models are tabulated in Table 7.2

TABLE 7.2
Calculated Values for all the Designed Representations

	Model 1.0	Model 2.0	Model 3.0	Model 4.0
Q_{actual} fin	6.2351W	6.6662W	8.3329W	7.2596W
$Q_{without}$ fin	0.1168W	0.1751W	0.1751W	0.24865W
Q_{ideal} fin	10.392W	11.028W	13.071W	10.2609W
ϵ	53.40	38.0621	47.5785	29.1960
η	59.99 %	60.45%	63.75%	70.75%

7.8 CONCLUSION

By using numerical simulation built in COMSOL Multiphysics software, the analysis is successfully executed in the heat transfer domain and fluid flow conditions for different fin arrangements of an automotive radiator. The analysis of the variations of the velocity and the temperature in the direction where the coolant is flowing with the air flow and with the radiator fins are presented here in order to have a comparative data analysis on the basis of effectiveness and efficiency. The radiator with additional adjoint mid fins i.e., circular and squared fins will lead to further heating of the radiator yet will also enhance the efficiency with the highest in the case of circular mid fin attachment.

REFERENCES

[1] Abdullah H. Alessa, A. M. Maqableh, and S. Ammourah, "Enhancement of natural convection heat transfer from a fin by rectangular perforations with aspect ratio of two," *International Journal of Physical Sciences*, vol. 4, no. 10, pp. 540–547, October, 2009.

[2] M. S. Sohal, and J. E. O'Brien, "Improving air cooled condenser performance using winglets and oval tubes in a geothermal power plant," *Geothermal Resources Council Transactions*, vol. 25, pp. 1–7, 2001.

[3] K. W. Song, L. B. Wang, J. F. Fan, Y. H. Zhang, and S. Liu, "Numerical study of heat transfer enhancement of finned flat tube bank fin with vortex generators mounted on both surfaces of the fin," *Heat and Mass Transfer*, vol. 44, pp. 959–967, 2007.

[4] J. He, L. Liu, and A. M. Jacobi, "Air-Side Heat- Transfer Enhancement by a New Winglet-Type Vortex Generator Array in a Plain-Fin Round-Tube Heat Exchanger," *Journal of Heat Transfer*, vol. 132, pp. 1–9, 2010.

[5] Pradip B. Ramdas, and Dinesh K. kumar, "A study on the Heat Transfer Enhancement for Air Flow through a Duct with various Rib Inserts," *International Journal of Latest Trends in Engineering And Technology*, vol. 2, no. 4, pp. 479–485, 2013.

[6] J. P. Yadav and B. R. Singh, "Study on Performance Evaluation of Automotive Radiator," *S-JPSET*, vol. 2, no. 2, 2011.

[7] J. R. Patel, and A. M. Mavani, "Review paper on CFD Analysis of Automobile Radiator to improve its Thermal Analysis of Automobile radiator," *IJSRD*, 2014.

[8] J. D. Chung, B. K. Park, and J. S. Lee, "The combined effects of angle of attack and louver angle of a winglet pair on heat transfer enhancement," *International Journal of enhanced Heat Transfer*, vol. 10, pp. 31–43, 2003.

[9] K. M. Kwak, K. Torii, and K. Nishino, "Heat transfer and flow characteristics of fin-tube bundles with and without winglet-type vortex generators," 2002, Springer, Vol. 33, pp. 696–702.

[10] S. Yadav, and K. M. Pandey, "A Comparative Thermal Analysis of Pin Fins for Improved Heat Transfer in Forced Convection," *International conference on Processing of Materials, Minerals and Energy*, vol. 5, no. 1, Part 1, pp. 1711–1717, July 2016.

[11] S. Yadav, and K. M. Pandey, "A Parametric Thermal Analysis of Triangular Fins for Improved Heat transfer In Forced Convection," *Strojniški vestnik – Journal of Mechanical Engineering*, vol. 64, no. 6, pp. 401–411, 2018.

[12] www.enginebasics.com/Engine%20Basics%20Root%20Folder/Engine%20Cooling%20Pg5.html.

[13] www.engineeringtoolbox.com/convective-heat-transfer-d_430.

[14] www.theengineeringpost.com.

[15] www.wikipedia.com.

[16] www.comsol.com-multiphysics-finite-element-method.

[17] Ambuj Gupta, and Ayushi Gaur, "Analysis of heat transfer phenomena from different fin Geometries using CFD simulation in ANSYS," *IJESC*, vol. 9, no. 9, 2019, ISSN 2321 3361.

[18] P. Natarajan, "Heat Transfer Analysis in Light Passenger Car Radiator With Various Geometrical Configurations," *International Research Journal of Engineering and Technology (IRJET)*, vol. 05, no. 06, June 2018, e-ISSN: 2395-0056.

[19] doc.comsol.com/5.5/doc/com.comsol.help.comsol/comsol_ref_fluidflow.20.32.html.

8 An Environment-Friendly Emerging Technique for Machining
Minimum Quantity Lubrication

Roshan Lal Virdi, Amrit Pal, Sukhpal Singh Chatha, and Hazoor Singh Sidhu

CONTENTS

8.1 INTRODUCTION

The term "manufacturing" refers to the method of producing a product. Various manufacturing processes shape metals into useful shapes. The metal-cutting operation is the method of removing any unnecessary material from a workpiece using a cutting tool in order to achieve the desired form. Machining tools have been around for a long time period and have played a significant role in the manufacturing revolution's progress [1]. Ingarao, et al. [2] stressed that modern industry considers developing machining processes in accordance with cleaner production standards to be a critical problem. These requirements must be met while maintaining production rates, product quality, and avoiding added costs.

The developing interest for higher efficiency, item quality and by and large economy in assembling by machining, especially to address the difficulties tossed by advancement and worldwide expense intensity, demands high material removal rate with longer tool life. However, P. Leskover [3] described that high production rate machining with high machining speed, depth of cut, and feed rate are necessarily correlated with high heat generation and cutting temperature. Hard material

DOI: 10.1201/9781003242291-8

machining produces a lot of heat in the machining field. The heat generated at the machining area raises the machining zone's temperature, reducing the cutting tool's life. It also has a detrimental influence on the material of the workpiece by creating residual stresses. As a result of these high residual stresses, sub-surface small cracks form, making the work material more susceptible to corrosion and oxidation [4].

Cutting fluids is the common technique used worldwide in metal machining processes to increase tool life, workpiece size precision, and surface quality. Cutting fluids are coolants and lubricants that, when utilized correctly, can boost efficiency by permitting machining operations at high cutting parameters and increasing the production rate [5]. Major types of cutting fluids used for cooling-lubrication purposes are based on mineral oils. Moreover, chemicals like sulphur, chlorine and phosphorus, etc. are added to these oils to further enhance their lubricating properties. These chemicals are harmful to the environment. Cutting fluids are made from mineral oils that can pollute the atmosphere if improperly treated, causing harm to soil and water supplies. Operators are also influenced by the negative effects of machining fluids on the shop floor, such as skin and respiratory issues. To comply with strict environmental laws, cutting fluids must be treated properly. Many studies have attempted to decrease the use of machining fluids because of their harmful effects [6]. Cutting fluids can be more costly than cutting equipment in some machining operations. Any corporation would benefit from a reduction in the use of machining fluids [7]. The most general technique of applying machining fluids is to flood the cutting tool and workpiece zone. Cutting fluid consumption is high, normally several litres per min., resulting in substantial increases in handling, disposal, and purchase costs.

Cutting fluids are utilized to decrease friction by lubrication, which is a tribological phenomenon that occurs at the machining zone in the cutting operation. Cutting fluids are often utilized to eliminate heat from the machining area. Cutting fluids also help to wash the chips out of the machining environment [8,9].

Cutting fluids have a lot of benefits, but they also have numerous disadvantages. The presence of microbial masses, especially endo-toxin, on the working floor is caused by bacterial growth in machining fluids. Coolants used in the past were non-biodegradable and contaminated the atmosphere [10].

With sticker rules and regulations in place to protect the environment and workers' health, attempts are being made to develop new lubrication strategies in metal-cutting operations that can increase lubricant tribological efficiency without the use of toxic additives. Researchers looked at alternatives such as dry and the MQL technique [11] in order to establish environmentally friendly cutting fluids [12].

Dry cutting without the use of coolants, is an option for lowering machining costs. In dry-cutting operations, the friction between the contacting pairs is strong, causing a rise in temperature, which increases tool wear and dimensional inaccuracies. Dry cutting rates are only suitable at low cutting speeds, resulting in a low output rate to extend tool life [13]. Moreover, chip forming, which can lead to oxidation of the machined surface, cannot be washed away. Furthermore, difficult-to-machine materials produce high temperatures in the machining zone, which must be managed. Therefore, it is not possible to exclude cutting fluid from certain machining processes [14].

Moreover, the hard-to-cut materials play an important role in many manufacturing industries due to the high demand for high-tech mechanical parts [15]. Therefore, the cutting fluids have the ability to facilitate superior cooling-lubricating characteristics during the metal-cutting operation of hard-to-cut alloys and also should enhance the sustainability in machining [16]. In dry machining, no type of cutting fluid is utilized in the operation. Due to environmental and economical points of view, the metal-cutting operation is performed without any use of metal-working fluid but it has some disadvantages. High heat is generated during dry cutting which leads to excessive tool wear, resulting in decreasing the life of the cutting tool. Moreover, the chips generated during the dry cutting cannot be washed away from the cutting area and these chips cause worse surface quality [17,18]. The results in excessive tool wear rate and high cutting zone temperature during dry cutting showed that a dry environment was unsuitable for the cutting of hard materials at medium to high cutting speed [19]. Therefore, eliminating cutting fluid in some metal-cutting operations is not possible [14]. Moreover, it is well known that the use of machining fluid can enhance the surface finish, cutting tool life and reduce the cutting stresses [10]. So, another possible way to reduce the use of machining fluid is cryogenic cooling and MQL.

The implementation of MQL technology will mitigate the limitations of dry operations. MQL is defined as a mist of a little amount of cutting fluid with highly compressed air, directed towards the machining area, minimizing the high flow rates of flood fluid application. MQL refers to the method of using the smallest quantity of lubricant fluid available in a machining operation [20].

8.2   MINIMUM QUANTITY LUBRICATION (MQL) TECHNIQUE

Currently, researchers have shown great interest en route MQL strategy. The substitute for conventional cooling and dry machining is MQL machining. The MQL technique is implemented as an eco-friendly and economically favourable method [21]. Moreover, the MQL technique is the best alternative method for precisely delivering a small quantity of cooling fluid to the machining zone and meeting the requirements of being environmentally friendly [22].

MQL is an emerging technology in which very little amount of metal-working fluid, i.e., 10–500 ml/h is delivered with high pressurized air at the cutting area. The lubricating oil used in MQL provided the requisite lubrication, and the high-pressure mist effectively removed the chips. MQL can be applied to grinding, turning, milling, drillings, etc. The line diagram of MQL setup for drilling operation is presented in Figure 8.1 [23]. MQL has been tested by a number of researchers in a variety of machining methods, and the findings indicate that it can help solve numerous problems related to the use of cutting fluids, such as high machining costs and worker health issues [24]. As a result, MQL systems are implemented as an eco-friendly and cost-effective process [21].

MQL not only reduces the quantity of fluid in metal cutting but can also enhance the cooling-lubricating characteristics. In addition, developing superior lubricants, such as biodegradable or chlorine-free oils for MQL applications, reduces the environmental effect of conventional oils usage and enhances the machining performance [25]. In this method, aerosol form of high pressurized air and lubricant offers both cooling and lubricating effect along with reducing the adhesion tendency of work material by decreasing the friction between the contacting pairs.

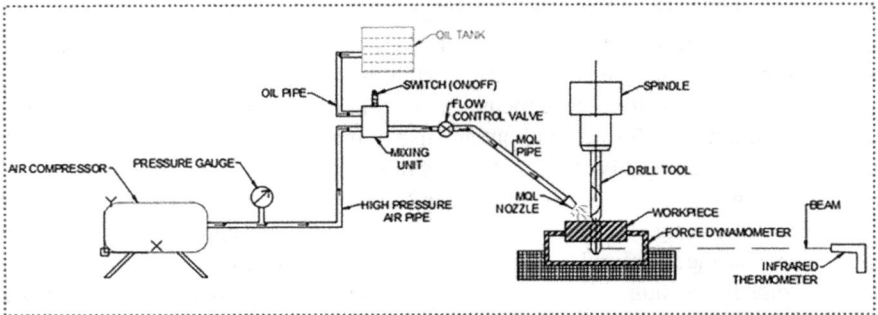

FIGURE 8.1   Line diagram of MQL setup for drilling [23].

Highly pressurized air confirms the cooling effect and oil offers the lubricating effect [26].

For superior efficiency of the MQL process, it must be confirmed that the lubricant mist must efficiently cover the contact zone, i.e., tool and work interfaces. MQL input parameters for example nozzle angle, spray velocity, size of droplet and distance between nozzle and contacting zone are to be selected for superior performance of MQL technology. Additionally, wettability and thermochemical properties of the cutting fluid can also affect the MQL cutting performance [27]. As suggested by Liao and Lin [28], the protective layer is developed at the machining interface. This is mainly because the extra oxygen is supplied by the MQL strategy. This protective layer performs as a diffusion barrier, which helps in enhancing the process performance and retaining the strength of the cutting tool by wear resistance. Many investigations have shown its benefits in enhancing the performance parameters relative to dry and conventional machining strategies [29].

Yuan, et al. studied the machining of Ti-6Al-4V using various cooling methods, including dry, flood, and MQL and found that MQL improved cutting stresses, wear rate, and surface quality [30]. Dhar et al. [31] discussed the impact of lubricant on wear rate and surface finish when machining AISI 4340 steel. The cutting efficiency of MQL machining was found to be superior to that of dry machining. Tool wear was minimized, and the surface finish was improved under the MQL jet. Tool wear can be minimized, improving tool life or increasing efficiency by allowing for higher cutting velocity and feed.

In another research work, Dhar, et al. [32] found that MQL is a viable substitute for turning in terms of surface quality, wear rate and heat generation in another analysis. MQL is superior to dry machining because it lowers the cutting temperature, lowers cutting forces and increases tool life. Using the MQL technique to further cool the machining zone during difficult-to-machine materials is still a challenge. Priarone, et al. [33] observed that in a milling operation, the MQL condition improved the tool life due to the good lubricating properties of the cutting fluid. Under the flood cooling condition, the tool wear reduces due to the higher cooling effect of the emulsion during the turning process. On the other hand, MQL has given the best surface quality results as compared to other cooling environments in turning.

Moreover, Fratila and Caizar [34] observed that the MQL method could be utilized instead of flood milling without influencing the milling characteristics. Hassanpour, et al. [35] examined that MQL milling of AISI 4340 significantly improves the milling performance parameters in terms of micro hardness, surface defects and roughness. Kang et al. [36] also reported that MQL technology demonstrated superior cutting performance during the milling at high milling velocity and observed the lowest tool wear among all the lubricating environments, i.e., MQL, conventional and dry.

Perçin, et al. [37] examined the influence of dry, conventional flood cooling, MQL, and cryogenic on the drilling characteristics in micro drilling of titanium alloy. It was reported that the magnitude of drilling forces considerably decreased under MQL environments. This is because the friction between the contacting surfaces is significantly reduced, thereby improving the surface quality of machined surfaces. Julia Dosbaeva [38] also described significant results of the MQL strategy during the machining of aluminium but advised focusing on improving the MQL setup for better cooling and lubricating the machining area. In another research work, Zeilmann and Weingaertner [39] also compared the machining performance with respect to cutting zone temperature during the drilling of titanium alloy under various MQL environments. It was revealed that internally supplied MQL oil could significantly reduce the drilling temperature as compared to externally applied MQL oil.

In order to make metal cutting more ecological, the MQL technique has been accepted as a successful near-dry machining application because of its eco-friendly capabilities. According to the metal-cutting process, reducing or eliminating the use of machining fluids is not possible. In these situations, it is very compulsory to develop alternative ways in order to avoid health and environment-related issues. Currently, MQL cutting with vegetable oil-based cutting fluids (VBCFs) as the base cutting fluid offered favourable findings in terms of cost and environmental-related issues. It also reduces the human health-related issues produced by petroleum oil-based cutting oils [40]. The use of vegetable oils in machining may permit this mixture to make possible the development of a new creation of machining fluid, where high machining performance along with good environmental compatibility could be attained [41].

8.3 VEGETABLE OIL-BASED MINIMUM QUANTITY LUBRICATION (MQL) MACHINING

Vegetable oil-based machining fluids are more advantageous in comparison to petroleum oil-based lubricants. These types of cutting oils are biodegradable, non-toxic, and long fatty acids in nature, and economically good relative to other cutting oils [42,43]. VBCFs have superior performance relative to petroleum oil-based cutting lubricants with respect to surface quality, wear rate and tool life [44,45]. The better machining performance of vegetable oils is mainly due to their good lubricating effect, high viscosity and high flashing point, and low evaporative losses in comparison to petroleum-based oils [46]. Also, the accessibility of mineral oil-based cutting oils is limited because of their fewer resources and they are decreasing very fast, on the other hand, vegetable oils are sustainable and renewable. Because vegetable oils are formulated from renewable sources and thus, they are unlimited and more sustainable [47]. Therefore, VBCFs may be selected as a better alternative to petroleum oil-based cutting oils. Recently, MQL machining with VBCFs presents many promising applications. In addition, the technical and industrial sector has already recognized that vegetable oils are a good substitute for mineral-based oils, as described by Li, et al. [48]. Furthermore, MQL with vegetable oils as base oil presents excellent results during the machining of hard-to-cut alloys.

Actually, the presence of long polar groups (–COOH and –COOR) in vegetable oils offers superior lubricating characteristics during cutting operation [49]. Because vegetable oils mainly consist of tri-glycerides which are glycerol molecules with long chain fatty acids attached at the hydroxyl groups through ester linkages. Long chains of polar fatty acids give high stability to thin lubricant layers. This layer strongly interacts with work material surfaces and reduces both friction and wear [50]. The schematic representation of the lubricating effect between contacting surfaces during vegetable oil-based MQL drilling environments is shown in Figure 8.2 [51]. Most vegetable oils contain at least four and sometimes as many as twelve different fatty acids. The percentage of each fatty acid depends not only on the type of the plant but also on the geo-climate and the weather available. In general, vegetable oils have poor oxidative and thermal stability when compared to mineral-based oils [52].

In addition, viscosity and heat transmission rate are extremely significant characteristics of the cutting oils. The friction and heat generation in the machining area is mainly affected by the viscosity and heat transmission rate of the cutting oil [53,54].

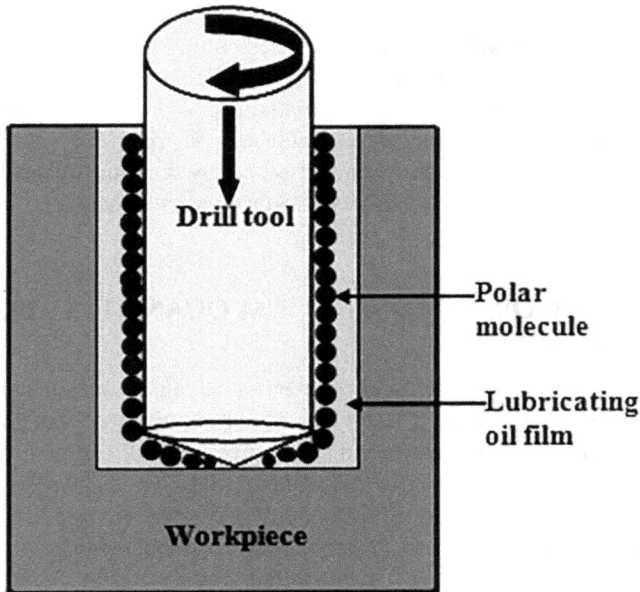

FIGURE 8.2   The schematic representation of the lubricating effect between contacting surfaces during vegetable oil-based MQL drilling environments [51].

Moreover, vegetable oils are more able to maintain their viscosity at a high machining temperature range relative to mineral-based cutting fluids [55]. As the temperature decreases, plant-based oils retain more oil than mineral-based oils in the machining zone and quickly wash out metal chips from the cutting area. Therefore, the high viscosity of vegetable oils confirms that vegetable oils offer a better lubricating effect [56]. In addition, the size of molecules in vegetable oils is homogenous relative to mineral-based oils, as a result of which the main properties, i.e., viscosity, flash and boiling point of vegetable oils are more stable [57].

A few years ago, Belluco and Chiffre [58] evaluated the performance of five vegetable oil-based cutting oils during the drilling of AISI 316L. The results indicated that drill life was significantly enhanced with the application of vegetable oil in comparison to dry drilling. Moreover, Ojolo, et al. [59] evaluated the lubricating characteristics of some plant-based oils during the machining of mild steel, aluminium and copper. Rahim and Sasahara [60] studied that vegetable oil, i.e., palm oil significantly enhances the lubrication effect during the drilling process. The results showed that palm oil presented superior performance with respect to drilling performance. Emami, et al. [61] described the influence of various types of cutting oils such as mineral, vegetable, and synthetic oils during MQL grinding. Results revealed that MQL technology directly affects the grinding performance.

Additionally, the utilization of plant-based cutting oils in the MQL method decreases environmental issues. Wang, et al. [62] studied the cooling-lubricating performance of different vegetable oil under flood and MQL grinding strategies. Findings indicated that MQL grinding with vegetable oil-based machining oils has excellent cooling lubricating characteristics in comparison to MQL with mineral oil and flood conditions. This is because of the high binding energy and low friction coefficient of vegetable oil-based cutting oils. Furthermore, Burton et al. [63] observed that the burr formation and chip thickness considerably reduced with the application of water emulsion and canola oil as compared to flood fluid. Lawal et al. [64] also revealed that machining with vegetable oils is a better alternative to petroleum oil-based cutting oils, and it presents superior process performance. Zhang et al. [44] considered 45 steel as workpiece during MQL grinding under various vegetable oils and reported better performance of vegetable oils concerning grinding energy, coefficient of friction and temperature.

The cooling ability of the MQL system is critical to its performance. As a result, the task is to improve the heat-carrying ability or thermal conductivity of machining lubricants. Heat-carrying entities can be introduced into lubricants to solve these problems. Nanofluid may be the key to these problems. It is described as a mixture of solid and liquid suspension. These are liquids made by mixing nano-sized particles into a base oil such as water, engine oil, vegetable and cutting oils, and so on. Many experiments have shown that mixing nano-solid particles of high conductivity into the base fluid enhances the heat transmission of the fluid [65,66].

Solid particles such as Al_2O_3, MoS_2, CNT, CuO, SiO_2, ZrO_2, and others are widely used in nanofluids. As these oxides and pure carbon molecules are applied to the base fluid in MQL machining, the fluid's heat-carrying and load-bearing capacity is increased, which improves the machining operation [67].

The heat transmission of liquids containing suspended nano solid particles such as copper into a base oil improved cooling effects by more than two times [68]. Nanofluid consisting of copper nanoparticles mixed in ethylene glycol shows an improved heat transmission rate relative to either pure ethylene glycol [69]. Nanofluids' properties make them appealing for cooling and lubricating applications in a variety of industries, especially in manufacturing processes.

The conventional method of lubrication in machining by using mineral oil-based cutting fluids poses an environmental risk. Since traditional fluids are often flammable, the contaminant waste that comes from their disposal from the finished product includes dangerous elements such as arsenic, sulphur, chlorine, and zinc [70].

As a consequence, it's important to pay attention to environmentally friendly lubricants and to consider the value of using biodegradable lubricants. As a result, biodegradable cutting fluids must be used instead of petroleum-based metal machining fluids. Vegetable oils are biodegradable and possess the required properties of cooling-lubrication. However, their physiochemical properties governed the cooling-lubrication phenomenon. Researchers are focusing on environment-friendly cutting fluids. Zareh-Desari and Davoodi [71] reported that vegetable oils enriched with nanoparticles decrease the friction between the contacting pairs, and resulted in better lubricity during the metal forming process.

The diffusion characteristics of oils play a vital role during cutting operations. Viscosity and wettability are the major properties that govern the cooling-lubrication phenomenon in machining processes. More viscous vegetable oils produced a stable film on the machining zone that enhances the cooling-lubrication phenomenon by lowering the friction. Furthermore, better wettability can be achieved by lowering the contact angle. For the sake of improving the cooling-lubrication phenomenon, researchers proposed the addition of nanoparticles in the base oil [72].

8.4 VEGETABLE OIL-BASED NANOFLUID MINIMUM QUANTITY LUBRICATION (NFMQL) MACHINING

While the main aim of MQL technology is to reduce the amount of coolant, on the other side various research scholars are trying to enhance the capabilities of machining fluid by using the nano-cutting fluids [73]. Recently, nano-cutting fluids are developed in order to boost the lubricating stability and thermal transmission capability during cutting operations. Nanoparticles of average size (<100 nm) mixed into the base fluid are known as nano-cutting fluids. It has been examined that nano lubricants play a main role in changing heat transport phenomena and the physical stability of machining oils to a greater extent. Various nanoparticles have been added for improving the efficiency of cutting oils, this includes metal and ceramic-based nanoparticles. These include several nanoparticles such as MoS_2, SiO_2, graphene, diamond and Al_2O_3, and are mainly utilized to improve the thermal and cooling- lubricating characteristics of metalworking fluids [74].

Nanofluids are fluids that are made by mixing metal oxide nanoparticles with lubricating oil. To keep particles from clumping together, dispersants and ultrasonic vibrations are used. One of the difficulties in the preparation of nanofluid is the dispersion. The Van der Waals force causes the nanoparticles to clump together and precipitate. The stability of nanoparticles mixed in base fluid without sedimentation or precipitation because of the downward-force of the accumulated weight is referred to as nanofluid dispersibility.

Nanoparticles enhance tribological performance considerably by improving the heat transmission rate of the coolant [75]. The effectiveness of nano lubricants depends upon various factors like the structure of the nanoparticle, morphology, size of the nanoparticle and nanoparticles concentration or quantity, and how the nano-cutting oil is supplied into the cutting area [76]. Plant oil-based nano-cutting fluids offer outstanding process performance by enhancing machining characteristics. Moreover, nanoparticles form a thin layer between the contacting pairs, thereby performing an anti-wear mechanism, which decreases wear rate and friction [77]. Nanoparticles possess improved thermo-physical characteristics because of their large surface-to-volume ratio, when mixed in small amounts by volume in vegetable oil, resulting in formulations which imbibe the environment-friendly and user-friendly features of the base oil and do not affect the environment like traditional cooling fluids [78].

According to the fact that the formulation of nanofluids is rather expensive, these new suspensions are not desirable for utilization in traditional flood machining. However, because the MQL strategy requires a less amount of oil, the application of nano-cutting oil in this technique is more suitable.

Viscosity is a significant factor of inner friction and flow resistance of a fluid. Wu et al. [79] described that the viscosity of nano-cutting fluid is higher than the viscosity of cutting oil without mixing nanoparticles. The high kinematic-viscosity of nanofluid results in the development of a layer between the contacting surfaces, which efficiently reduces the friction and enhances the cooling-lubricating capabilities.

In addition, the Sommerfeld number increases as the viscosity of nano-cutting fluid increases [80], due to the lubrication region converting from boundary to film lubrication. Thus, the coefficient of friction of nano-cutting fluid is lower than that pure flood lubrication. As well as the heat transmission rate of nano-cutting oil is higher than that of pure fluid [81] and [82]. This is mainly due to the suspended nanoparticles increasing the heat transmission rate of the base oil [83]. Figure 8.3 illustrates a schematic representation of the lubricating mechanism with the application of nano-cutting fluids between the mating surfaces [23].

Various research studies have been conducted to explore the process characteristics during NFMQL machining. Many investigations are based on the mixing of nanoparticles in the base fluid and its evaluation [84]. Chatha, et al. [77] investigated that the drilling enhancement in terms of drilling stresses, hole quality and tool life significantly enhances during NFMQL drilling relative to other cooling-lubricating methods. This is mainly because of the roll-bearing effect and superior cooling characteristics of nano-cutting oils. Nam, et al. [85] examined the process performance during nanofluid MQL drilling of titanium alloy. Results indicated that higher wt.% of nanoparticles notably decrease the cutting forces and edge-radius at lower drilling parameters. This is because of high wt.% of nano lubricants could perform bearing and sliding effects.

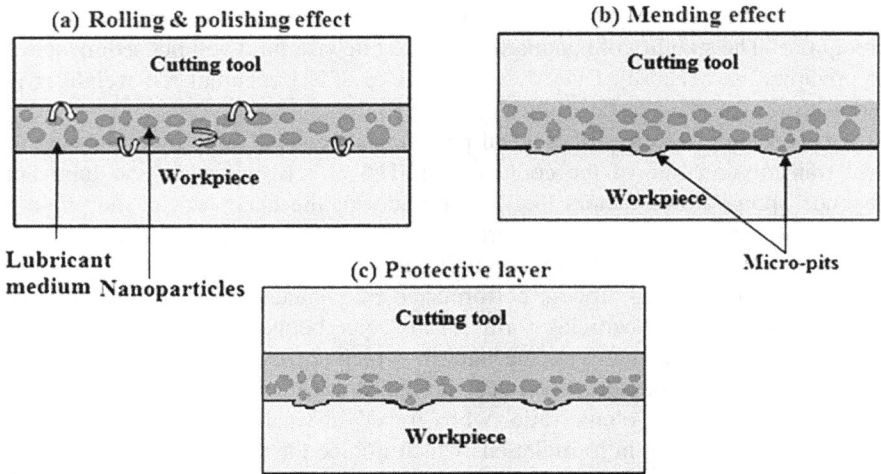

FIGURE 8.3   A schematic representation of the lubricating mechanism with the application of nano-cutting fluids between the mating surfaces [23].

Amrita, et al. [86] investigated the performance of nano graphene-based cutting oil in machining and revealed that the MQL turning enhanced the machining performance relative to flood cooling strategy by reducing the turning stresses, machining temperature, roughness and wear rate, along with enhancing that chip morphology. In addition, Park, et al. [87] investigated the influence of nano oil during MQL milling and examined superior performance, particularly in tool wear. Najiha, et al. [88] studied the influence of nano-TiO2-based oil during MQL milling with respect to wear rate. It was observed that the high concentration (2.5%) of TiO_2 reduced the wear rate during NFMQL milling due to the nano-bearing effect.

Mao, et al. [89] examined the grinding characteristics of AISI-52100 steel and reported that Al_2O_3 nanofluid MQL significantly reduced the forces, friction and temperature. Emami, et al. [61] investigated the influence of nano-Al_2O_3 using various oils viz. petroleum, hydro-cracked, synthetic and vegetable oil-based machining fluids to examine the grinding performance. It was revealed that plant oil-based MQL grinding is better and eco-friendly relative to other cooling environments. To save the environment, researchers are trying to replace traditional non-biodegradable oils.

Thus, this paper focuses on the way to develop the MQL technology more efficiently, sustainable and eco-friendly, when machining. In this review study, the important key to enhancing the cooling-lubricating performance of the MQL method is using special metalworking fluids, such as biodegradable oils and nano-cutting oils, which could enhance the wettability capabilities of the base oil.

8.5   CONCLUSIONS

Researchers found that during the grinding process, most of the energy is wasted in the form of heat which is produced due to the friction between the work material and the rotating grinding wheel and also concluded that the heat generated adversely affects the process outcome. Therefore, the need for lubrication arises to dissipate the localized heat. Lubricants help in minimizing these adverse effects and help in increasing tool life and surface finish. During lubrication when the lubricant is applied it forms a sort of film between the rubbing surfaces, which prevents the direct contact between the two mating surfaces. As a result, the cooling and lubrication efficiency of the grinding fluid is a critical technical aspect of the MQL grinding process. Cutting fluids made from vegetables are environmentally friendly and have the same machining efficiency. Bio-based cutting fluids used in the nanofluid MQL device are thought to reduce the health and environmental risks associated with traditional coolants. In a nutshell, for green machining, it is essential to concentrate on a bio-based MQL system.

REFERENCES

[1] P. N. Rao, *Manufacturing Technology: Metal Cutting and Machine Tools.* New Delhi: Tata McGraw-Hill Publishing Co. Ltd., 2005.

[2] G. Ingarao, P. C. Priarone, F. Gagliardi, R. Di Lorenzo, and L. Settineri, "Subtractive versus mass conserving metal shaping technologies: an environmental impact comparison," *Journal of Cleaner Production*, vol. 87, pp. 862–873, 2015/01/15/ 2015.

[3] J. G. P. Leskover, "The metallurgical aspect of machining," *Journal of Cleaner Production*, vol. Annales of CIRP 35, no. 1, pp. 537–550, 1986.

[4] E. B. H. K. Tonshoff, "Determination of the mechanical and thermal influences on machined surface by microhardness and residual stress analysis," *Journal of Cleaner Production*, vol. 87, pp. 519–532, 1986.

[5] J. Beddoes and M. J. Bibby, *Principles of Metal Manufacturing Processes.* Burlington, MA: Butterworth-Heinemann, 2009.

[6] M. Soković and K. Mijanović, "Ecological aspects of the cutting fluids and its influence on quantifiable parameters of the cutting processes," *Journal of Materials Processing Technology*, vol. 109, pp. 181–189, 2/1/ 2001.

[7] G. E. F. Klocke, "Dry cutting," *CIRP*, vol. 46, no. 2, pp. 519–526, 1997.

[8] Z. Pawlak, B. E. Klamecki, T. Rauckyte, G. P. Shpenkov, and A. Kopkowski, "The tribochemical and micellar aspects of cutting fluids," *Tribology International*, vol. 38, pp. 1–4, 1// 2005.

[9] S. Ebbrell, N. H. Woolley, Y. D. Tridimas, D. R. Allanson, and W. B. Rowe, "The effects of cutting fluid application methods on the grinding process," *International Journal of Machine Tools and Manufacture*, vol. 40, pp. 209–223, 1// 2000.

[10] A. Shokrani, V. Dhokia, and S. T. Newman, "Environmentally conscious machining of difficult-to-machine materials with regard to cutting fluids," *International Journal of Machine Tools and Manufacture*, vol. 57, pp. 83–101, 6// 2012.

[11] S. Min, I. Inasaki, S. Fujimura, T. Wada, S. Suda, and T. Wakabayashi, "A Study on Tribology in Minimal Quantity Lubrication Cutting," *CIRP Annals*, vol. 54, pp. 105–108, 2005/01/01/ 2005.

[12] V. Kumar, S. K. Sinha, and A. K. Agarwal, "Tribological Studies of an Internal Combustion Engine," in *Advanced Engine Diagnostics*, A. K. Agarwal, J. G. Gupta, N. Sharma, and A. P. Singh, Eds., ed Singapore: Springer Singapore, 2019, pp. 237–253.

[13] M. P. Groover, *Fundamentals of Modern Manufacturing.* United State.: John Wiley & Sons, 2002.

[14] S. M. Alves and J. F. G. de Oliveira, "Development of new cutting fluid for grinding process adjusting mechanical performance and environmental impact," *Journal of Materials Processing Technology*, vol. 179, pp. 185–189, 10/20/ 2006.

[15] A. Khatri, M. P. Jahan, and J. Ma, "Assessment of tool wear and microstructural alteration of the cutting tools in conventional and sustainable slot milling of Ti-6Al-4V alloy," *The International Journal of Advanced Manufacturing Technology*, vol. 105, pp. 2799–2814, 2019.

[16] P. Huang, H. Li, W.-L. Zhu, H. Wang, G. Zhang, X. Wu, *et al.*, "Effects of eco-friendly cooling strategy on machining performance in micro-scale diamond turning of Ti–6Al–4V," *Journal of Cleaner Production*, vol. 243, p. 118526, 2020.

[17] G. Krolczyk, R. Maruda, J. Krolczyk, S. Wojciechowski, M. Mia, P. Nieslony, *et al.*, "Ecological trends in machining as a key factor in sustainable production–a review," *Journal of Cleaner Production*, vol. 218, pp. 601–615, 2019.

[18] M. M. R. Nune and P. K. Chaganti, "Development, characterization, and evaluation of novel eco-friendly metal working fluid," *Measurement*, vol. 137, pp. 401–416, 2019.

[19] H. Sasahara, A. Kato, H. Nakajima, H. Yamamoto, T. Muraki, and M. Tsutsumi, "High-speed rotary cutting of difficult-to-cut materials on multitasking lathe," *International journal of machine tools and manufacture*, vol. 48, pp. 841–850, 2008.

[20] T. Obikawa, Y. Kamata, and J. Shinozuka, "High-speed grooving with applying MQL," *International Journal of Machine Tools and Manufacture*, vol. 46, pp. 1854–1861, 11// 2006.

[21] A. D. Jayal, A. K. Balaji, R. Sesek, A. Gaul, and D. R. Lillquist, "Machining Performance and Health Effects of Cutting Fluid Application in Drilling of A390.0 Cast Aluminum Alloy," *Journal of Manufacturing Processes*, vol. 9, pp. 137–146, // 2007.

[22] E. Vazquez, J. Gomar, J. Ciurana, and C. A. Rodríguez, "Analyzing effects of cooling and lubrication conditions in micromilling of Ti6Al4V," *Journal of Cleaner Production*, vol. 87, pp. 906–913, 2015.

[23] A. Pal, S. S. Chatha, and H. S. Sidhu, "Performance evaluation of the minimum quantity lubrication with Al2O3-mixed vegetable-oil-based cutting fluid in drilling of AISI 321 stainless steel," *Journal of Manufacturing Processes*, vol. 66, pp. 238–249, 2021.

[24] A. Attanasio, M. Gelfi, C. Giardini, and C. Remino, "Minimal quantity lubrication in turning: Effect on tool wear," *Wear*, vol. 260, pp. 333–338, 2/10/ 2006.

[25] M. Emami, M. Sadeghi, and A. A. Sarhan, "Investigating the effects of liquid atomization and delivery parameters of minimum quantity lubrication on the grinding process of Al2O3 engineering ceramics," *Journal of Manufacturing Processes*, vol. 15, pp. 374–388, 2013.

[26] F. Rabiei, A. Rahimi, M. Hadad, and M. Ashrafijou, "Performance improvement of minimum quantity lubrication (MQL) technique in surface grinding by modeling and optimization," *Journal of Cleaner Production*, vol. 86, pp. 447–460, 2015.

[27] K.-H. Park, J. Olortegui-Yume, M.-C. Yoon, and P. Kwon, "A study on droplets and their distribution for minimum quantity lubrication (MQL)," *International Journal of Machine Tools and Manufacture*, vol. 50, pp. 824–833, 2010.

[28] Y. Liao and H. Lin, "Mechanism of minimum quantity lubrication in high-speed milling of hardened steel," *International Journal of Machine Tools and Manufacture*, vol. 47, pp. 1660–1666, 2007.

[29] S. Chinchanikar and S. Choudhury, "Hard turning using HiPIMS-coated carbide tools: Wear behavior under dry and minimum quantity lubrication (MQL)," *Measurement*, vol. 55, pp. 536–548, 2014.

[30] S. M. Yuan, L. T. Yan, W. D. Liu, and Q. Liu, "Effects of cooling air temperature on cryogenic machining of Ti–6Al–4V alloy," *Journal of Materials Processing Technology*, vol. 211, pp. 356–362, 3/1/ 2011.

[31] N. R. Dhar, M. Kamruzzaman, and M. Ahmed, "Effect of minimum quantity lubrication (MQL) on tool wear and surface roughness in turning AISI-4340 steel," *Journal of Materials Processing Technology*, vol. 172, pp. 299–304, 2006.

[32] N. R. Dhar, M. T. Ahmed, and S. Islam, "An experimental investigation on effect of minimum quantity lubrication in machining AISI 1040 steel," *International Journal of Machine Tools and Manufacture*, vol. 47, pp. 748–753, 2007.

[33] P. C. Priarone, M. Robiglio, L. Settineri, and V. Tebaldo, "Milling and turning of titanium aluminides by using minimum quantity lubrication," *Procedia Cirp*, vol. 24, pp. 62–67, 2014.

[34] D. Fratila and C. Caizar, "Application of Taguchi method to selection of optimal lubrication and cutting conditions in face milling of AlMg3," *Journal of Cleaner Production*, vol. 19, pp. 640–645, 2011.

[35] H. Hassanpour, M. H. Sadeghi, A. Rasti, and S. Shajari, "Investigation of surface roughness, microhardness and white layer thickness in hard milling of AISI 4340 using minimum quantity lubrication," *Journal of Cleaner Production*, vol. 120, pp. 124–134, 2016.

[36] M. Kang, K. Kim, S. Shin, S. Jang, J. Park, and C. Kim, "Effect of the minimum quantity lubrication in high-speed end-milling of AISI D2 cold-worked die steel (62 HRC) by coated carbide tools," *Surface and Coatings Technology*, vol. 202, pp. 5621–5624, 2008.

[37] M. Perçin, K. Aslantas, I. Ucun, Y. Kaynak, and A. Cicek, "Micro-drilling of Ti–6Al–4V alloy: The effects of cooling/lubricating," *Precision engineering*, vol. 45, pp. 450–462, 2016.

[38] G. F.-R. Julia Dosbaeva, Jean Dasch, and Stephen Veldhuis, "Enhancement of Wet- and MQL-Based Machining of Automotive Alloys Using Cutting Tools with DLC/Polymer Surface Treatments," *Journal of Materials Engineering and Performance*, vol. 17, pp. 346–351, 2008.

[39] R. P. Zeilmann and W. L. Weingaertner, "Analysis of temperature during drilling of Ti6Al4V with minimal quantity of lubricant," *Journal of Materials Processing Technology*, vol. 179, pp. 124–127, 2006.

[40] S. A. Lawal, I. A. Choudhury, and Y. Nukman, "A critical assessment of lubrication techniques in machining processes: a case for minimum quantity lubrication using vegetable oil-based lubricant," *Journal of Cleaner Production*, vol. 41, pp. 210–221, 2013.

[41] S. M. Alves, B. S. Barros, M. F. Trajano, K. S. B. Ribeiro, and E. Moura, "Tribological behavior of vegetable oil-based lubricants with nanoparticles of oxides in boundary lubrication conditions," *Tribology international*, vol. 65, pp. 28–36, 2013.

[42] M. H. Cetin, B. Ozcelik, E. Kuram, and E. Demirbas, "Evaluation of vegetable based cutting fluids with extreme pressure and cutting parameters in turning of AISI 304L by Taguchi method," *Journal of Cleaner Production*, vol. 19, pp. 2049–2056, 2011.

[43] C.-M. Lee, Y.-H. Choi, J.-H. Ha, and W.-S. Woo, "Eco-friendly technology for recycling of cutting fluids and metal chips: A review," *International Journal of Precision Engineering and Manufacturing-Green Technology*, vol. 4, pp. 457–468, 2017.

[44] Y. Zhang, C. Li, D. Jia, D. Zhang, and X. Zhang, "Experimental evaluation of MoS2 nanoparticles in jet MQL grinding with different types of vegetable oil as base oil," *Journal of Cleaner Production*, vol. 87, pp. 930–940, 2015.

[45] Q. Yin, C. Li, L. Dong, X. Bai, Y. Zhang, M. Yang, *et al.*, "Effects of Physicochemical Properties of Different Base Oils on Friction Coefficient and Surface Roughness in MQL Milling AISI 1045," *International Journal of Precision Engineering and Manufacturing-Green Technology*, 2021/02/10 2021.

[46] S. A. Lawal, I. A. Choudhury, and Y. Nukman, "Evaluation of vegetable and mineral oil-in-water emulsion cutting fluids in turning AISI 4340 steel with coated carbide tools," *Journal of Cleaner Production*, vol. 66, pp. 610–618, 2014.

[47] K. K. Gajrani and M. R. Sankar, "Past and current status of eco-friendly vegetable oil based metal cutting fluids," *Materials Today: Proceedings*, vol. 4, pp. 3786–3795, 2017.

[48] M. Li, T. Yu, L. Yang, H. Li, R. Zhang, and W. Wang, "Parameter optimization during minimum quantity lubrication milling of TC4 alloy with graphene-dispersed vegetable-oil-based cutting fluid," *Journal of cleaner production*, vol. 209, pp. 1508–1522, 2019.

[49] S. Debnath, M. M. Reddy, and Q. S. Yi, "Environmental friendly cutting fluids and cooling techniques in machining: a review," *Journal of cleaner production*, vol. 83, pp. 33–47, 2014.

[50] N. Fox and G. Stachowiak, "Vegetable oil-based lubricants—a review of oxidation," *Tribology international*, vol. 40, pp. 1035–1046, 2007.

[51] A. Pal, S. S. Chatha, and H. S. Sidhu, "Performance Evaluation of Various Vegetable Oils and Distilled Water as Base Fluids Using Eco-friendly MQL Technique in Drilling of AISI 321 Stainless Steel," *International Journal of Precision Engineering and Manufacturing-Green Technology*, 2021/05/27 2021.

[52] A. Vaibhav Koushik, S. Narendra Shetty, and C. Ramprasad, "Vegetable oil-based metal working fluids-A review," *International Journal on Theoretical and applied Reasearch in Mechanical Engineering*, vol. 1, pp. 95–101, 2012.

[53] O. Fasina and Z. Colley, "Viscosity and specific heat of vegetable oils as a function of temperature: 35 C to 180 C," *International Journal of Food Properties*, vol. 11, pp. 738–746, 2008.

[54] A. Turgut, I. Tavman, and S. Tavman, "Measurement of thermal conductivity of edible oils using transient hot wire method," *International journal of food properties*, vol. 12, pp. 741–747, 2009.

[55] S. R, R. J. H. N, S. K. J, and G. M. Krolczyk, "A comprehensive review on research developments of vegetable-oil based cutting fluids for sustainable machining challenges," *Journal of Manufacturing Processes*, vol. 67, pp. 286–313, 2021/07/01/ 2021.

[56] E. Kuram, B. Ozcelik, and E. Demirbas, "Environmentally friendly machining: vegetable based cutting fluids," in *Green manufacturing processes and systems*, ed: Springer, 2013, pp. 23–47.

[57] K. Ulrich, "Vegetable oil-based coolants improve cutting performance," *Journal of cutting fluids*, htt://www. blaser. com/download/Dec02. pdf, 2002.

[58] W. Belluco and L. De Chiffre, "Performance evaluation of vegetable-based oils in drilling austenitic stainless steel," *Journal of materials processing technology*, vol. 148, pp. 171–176, 2004.

[59] S. Ojolo, M. Amuda, O. Ogunmola, and C. Ononiwu, "Experimental determination of the effect of some straight biological oils on cutting force during cylindrical turning," *Matéria (Rio de Janeiro)*, vol. 13, pp. 650–663, 2008.

[60] E. A. Rahim and H. Sasahara, "An analysis of surface integrity when drilling inconel 718 using palm oil and synthetic ester under MQL condition," *Machining Science and Technology*, vol. 15, pp. 76–90, 2011.

[61] M. Emami, M. H. Sadeghi, A. A. D. Sarhan, and F. Hasani, "Investigating the Minimum Quantity Lubrication in grinding of Al2O3 engineering ceramic," *Journal of Cleaner Production*, vol. 66, pp. 632–643, 2014.

[62] Y. Wang, C. Li, Y. Zhang, M. Yang, B. Li, D. Jia, *et al.*, "Experimental evaluation of the lubrication properties of the wheel/workpiece interface in minimum quantity lubrication (MQL) grinding using different types of vegetable oils," *Journal of Cleaner Production*, vol. 127, pp. 487–499, 2016.

[63] G. Burton, C.-S. Goo, Y. Zhang, and M. B. Jun, "Use of vegetable oil in water emulsion achieved through ultrasonic atomization as cutting fluids in micro-milling," *Journal of Manufacturing Processes*, vol. 16, pp. 405–413, 2014.

[64] S. Lawal, I. Choudhury, and Y. Nukman, "Application of vegetable oil-based metalworking fluids in machining ferrous metals—a review," *International Journal of Machine Tools and Manufacture*, vol. 52, pp. 1–12, 2012.

[65] P. Keblinski, S. R. Phillpot, S. U. S. Choi, and J. A. Eastman, "Mechanisms of heat flow in suspensions of nano-sized particles (nanofluids)," *International Journal of Heat and Mass Transfer*, vol. 45, pp. 855–863, 2// 2002.

[66] F. Lockwood, Z. Zhang, T. Forbus, and S. e. a. Choi, "The Current Development of Nanofluid Research," *SAE Technical Paper 2005-01-1929*, 2005.

[67] B. Li, C. Li, Y. Zhang, Y. Wang, D. Jia, M. Yang, *et al.*, "Heat transfer performance of MQL grinding with different nanofluids for Ni-based alloys using vegetable oil," *Journal of Cleaner Production*, vol. 154, pp. 1–11, 2017/06/15/ 2017.

[68] S. U. S. Choi, "Enhancing Thermal Conductivity of Fluids with Nano Particles, Developments and Applications of Non-Newtonian Flows," *FED*, vol. 231, pp. 99–105, 1995.

[69] S. U. S. Choi, Z. G. Zhang, W. Yu, F. E. Lockwood, and E. A. Grulke, "Anomalous thermal conductivity enhancement in nanotube suspensions," *Applied Physics Letters*, vol. 79, p. 2252, 2001.

[70] K. P. Rao and C. L. Xie, "A comparative study on the performance of boric acid with several conventional lubricants in metal forming processes," *Tribology International*, vol. 39, pp. 663–668, 2006/07/01/ 2006.

[71] B. Zareh-Desari and B. Davoodi, "Assessing the lubrication performance of vegetable oil-based nano-lubricants for environmentally conscious metal forming processes," *Journal of Cleaner Production*, vol. 135, pp. 1198–1209, 2016/11/01/ 2016.

[72] B. Shen, A. J. Shih, and S. C. Tung, "Application of Nanofluids in Minimum Quantity Lubrication Grinding," *Tribology Transactions*, vol. 51, pp. 730–737, 2008.

[73] C. Chan, W. Lee, and H. Wang, "Enhancement of surface finish using water-miscible nano-cutting fluid in ultra-precision turning," *International Journal of Machine Tools and Manufacture*, vol. 73, pp. 62–70, 2013.

[74] N. A. C. Sidik, S. Samion, J. Ghaderian, and M. N. A. W. M. Yazid, "Recent progress on the application of nanofluids in minimum quantity lubrication machining: A review," *International Journal of Heat and Mass Transfer*, vol. 108, pp. 79–89, 2017.

[75] J. S. Nam, P.-H. Lee, and S. W. Lee, "Experimental characterization of micro-drilling process using nanofluid minimum quantity lubrication," *International Journal of Machine Tools and Manufacture*, vol. 51, pp. 649–652, 2011.

[76] B. Rahmati, A. A. Sarhan, and M. Sayuti, "Morphology of surface generated by end milling AL6061-T6 using molybdenum disulfide (MoS2) nanolubrication in end milling machining," *Journal of Cleaner Production*, vol. 66, pp. 685–691, 2014.

[77] S. S. Chatha, A. Pal, and T. Singh, "Performance evaluation of aluminium 6063 drilling under the influence of nanofluid minimum quantity lubrication," *Journal of cleaner production*, vol. 137, pp. 537–545, 2016.

[78] R. Padmini, P. V. Krishna, and G. K. M. Rao, "Effectiveness of vegetable oil based nanofluids as potential cutting fluids in turning AISI 1040 steel," *Tribology International*, vol. 94, pp. 490–501, 2016.

[79] Y. Wu, W. Tsui, and T. Liu, "Experimental analysis of tribological properties of lubricating oils with nanoparticle additives," *Wear*, vol. 262, pp. 819–825, 2007.

[80] C. Mao, Y. Huang, X. Zhou, H. Gan, J. Zhang, and Z. Zhou, "The tribological properties of nanofluid used in minimum quantity lubrication grinding," *The International Journal of Advanced Manufacturing Technology*, vol. 71, pp. 1221–1228, 2014.

[81] P. Keblinski, S. Nayak, P. Zapol, and P. Ajayan, "Charge distribution and stability of charged carbon nanotubes," *Physical review letters*, vol. 89, p. 255503, 2002.

[82] Y. Xuan and W. Roetzel, "Conceptions for heat transfer correlation of nanofluids," *International Journal of heat and Mass transfer*, vol. 43, pp. 3701–3707, 2000.

[83] L. Godson, B. Raja, D. Mohan Lal, and S. Wongwises, "Enhancement of heat transfer using nanofluids—An overview," *Renewable and Sustainable Energy Reviews*, vol. 14, pp. 629–641, 2010.

[84] Y. Wang, C. Li, Y. Zhang, B. Li, M. Yang, X. Zhang, et al., "Comparative evaluation of the lubricating properties of vegetable-oil-based nanofluids between frictional test and grinding experiment," *Journal of Manufacturing Processes*, vol. 26, pp. 94–104, 2017.

[85] J. Nam, J. W. Kim, J. S. Kim, J. Lee, and S. W. Lee, "Parametric analysis and optimization of nanofluid minimum quantity lubrication micro-drilling process for titanium alloy (Ti-6Al-4V) using response surface methodology and desirability function," *Procedia Manufacturing*, vol. 26, pp. 403–414, 2018.

[86] M. Amrita, R. Srikant, and A. Sitaramaraju, "Performance evaluation of nanographite-based cutting fluid in machining process," *Materials and Manufacturing Processes*, vol. 29, pp. 600–605, 2014.

[87] K.-H. Park, B. Ewald, and P. Y. Kwon, "Effect of nano-enhanced lubricant in minimum quantity lubrication balling milling," *Journal of tribology*, vol. 133, 2011.

[88] M. Najiha, M. Rahman, and K. Kadirgama, "Performance of water-based TiO2 nanofluid during the minimum quantity lubrication machining of aluminium alloy, AA6061-T6," *Journal of cleaner production*, vol. 135, pp. 1623–1636, 2016.

[89] C. Mao, X. Tang, H. Zou, X. Huang, and Z. Zhou, "Investigation of grinding characteristic using nanofluid minimum quantity lubrication," *International Journal of Precision Engineering and Manufacturing*, vol. 13, pp. 1745–1752, 2012/10/01 2012.

9 Effect of Surface Topography and Roughness on the Wetting Characteristics of an Indigenously Developed Green Cutting Fluid (GCF)

Vimal Edachery, Sindhu Ravi, Aliya F. Badiuddin, and Abel Tomy
Department of Mechanical Engineering, Indian Institute of Science, Bengaluru, India

Suvin P. S.
Department of Mechanical Engineering, National Institute of Technology Karnataka, Suratkal, India

Satish V. Kailas
Department of Mechanical Engineering, Indian Institute of Science, Bengaluru, India

CONTENTS

DOI: 10.1201/9781003242291-9

9.1 INTRODUCTION

In recent years, cutting fluids have seen a growth in demand and production, owing to their wide range of uses in the machining industry. Cutting fluids are the type of lubricant that minimize friction, wear, and abrasion between metals that come in contact with one another. They operate by establishing a layer of lubricant and inducing wetness in surfaces that slide over one another. The topography of a surface determines how much lubrication it can hold onto. When metals contact, asperities and cavities on a surface enable fluid to collect between them, creating a constant supply of lubricant that aids in the removal of heat and chips [1,2]. Also, these fluids determine a high material removal rate (MRR), decrease tool wear, and aid in a variety of tasks such as grinding, drilling, stamping, surface finishing, and so on. By eliminating built-up heat, lowering coefficients of friction at tool-chip and tool-work interfaces, flushing away the chip, and avoiding the development of Built-up edges (BUEs), they improve tool service life and surface quality [3–6].

However, commercial cutting fluids (CCFs) are based on mineral oils (MOs) which have proven to be toxic, and they contribute significantly to multiple environmental concerns. There are several additives in a cutting fluid such as biocides and corrosion inhibitors and many others. These additives are non-renewable and extremely harmful to the environment. Cutting fluid toxicity is caused by the following factors:

1. The fluid that comes into contact with the cutting fluid during its use
2. The bacterial and/or fungal cultures that develop over time
3. The corrosion inhibitors and biocides are applied to stop corrosion and protect the machine against the growth of these living things.

These fluids do not disintegrate quickly, and improper disposal can lead to soil degradation and water toxicity. They also pose serious human health hazards, such as respiratory disorders, heart disease, and skin irritation, when subjected to prolonged exposure. MOs are sourced from fossil fuels, which are finite and non-renewable resources. Environmental studies have indicated that the present natural resource extraction is 1.7 times faster than nature's regeneration potential [7–14].

Cleaner, non-polluting practices must be used in the industrial sector if these negative consequences on the environment and human health are to be minimized. Machining operations are one of the production sectors where metal-cutting fluids may be used to approach the standard of sustainability [15–19]. This rise in environmental consciousness among industries has observed a paradigm shift of using eco-friendly biodegradable oil-based cutting fluids instead of MO-based cutting fluids, in machining processes. Indigenously developed sustainable-oil-based cutting fluids, which are an upcoming panacea for the hazards posed by synthetic cutting fluids, are functional and sustainable substitutes. They do not compromise on their innumerable applications in the manufacturing and machining industry [20]. In this regard, Suvin et al. [14] have synthesized a Green Cutting Fluid (GCF) by using materials that have low toxicity, are eco-friendly, and can degrade easily and carried out experiments to determine the storage stability of emulsions along with corrosion tests. They characterized the GCF such that its properties resembled the MO-based cutting fluid but were minorly susceptible to corrosion, possessed good thermal conductivity and specific heat, and had significantly less toxicity as compared to the mineral-oil-based cutting fluids.

The target of this work is to establish the performance efficiency of the afore-mentioned GCF in comparative analysis to MO-based CCF on various surfaces. To do so, experimental research has been done to determine how surface topography, particularly directionality, and roughness, affects the GCF's wetting characteristics. Comparative tests were done between 5% and 10% concentration of the GCF to determine a more favourable concentration to facilitate maximum wetting. For this, the GCF was examined on three different unidirectional roughness (roughness range 0.06 μm to 2.1 μm) created on materials, AA5052, Ti6Al4V, and EN31. The effect of parameters of surface topography and roughness on wetting is analyzed. The experiments were carried out using the above-mentioned materials as sub-strates, using a WYKO NT1100 profilometer to analyze the surface topography and Contact Angle Goniometer to obtain the contact angles. These efforts are directed towards establishing sufficient evidence to conclude that GCF is an effective replacement for synthetic CCFs.

9.1.1 CONCEPTUALIZATION OF WETTING

Any cutting fluid's wettability, which is a measurement of its Contact Angle (CA), determines how well it wets a surface. The angle created when a liquid-vapor interface meets a solid surface is known as the contact angle, and it is commonly measured through the liquid [21–23]. A convenient and simple approach to measuring contact angle is to find the tangent of a liquid drop with a solid surface at the base and measure the angle. Water's contact angle with a surface serves as a common reference point for comparing the contact angles of various liquids with the same surface. The interaction between solids and liquids is studied through experimental measurements of contact angle [24–26]. The compre-hensive research conducted by Thomas Young gave rise to Young's Equation which explains the wetting phenomenon in detail. Young's Equation, as represented in Equation 9.1, is the primary equation that relates contact angle θ, liquid surface tension γ_l, solid surface tension γ_s and solid-liquid surface tension γ_{sl}.

$$\gamma_l \cos \theta = \theta \gamma_s \gamma_{sl} \qquad (9.1)$$

In any case, perfect wetting requires a maximum value of $\cos\theta$. This condition is fulfilled when the contact angle θ is zero. From the equation, it can be observed that the measure of $\cos\theta$ can be increased in two ways: by lowering the surface tension

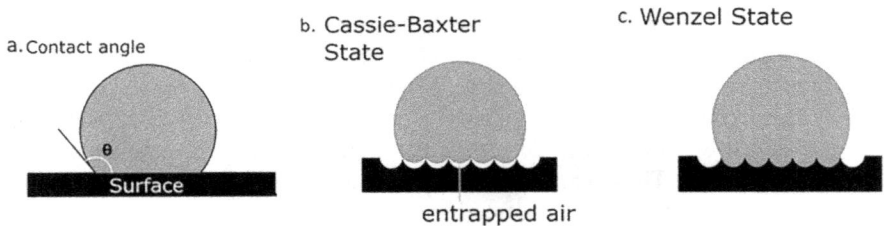

FIGURE 9.1 a) Representation of contact angle b) Cassie-Baxter state wetting c) Wenzel state wetting.

of the liquid or by raising the surface energy of the solid. These methods can be employed when it is required to increase the wetting properties of the liquid on a surface. Earlier researchers in their study on pillar-like surface topographies have also observed that wettability as a function of contact angle depends on the height of the pillar, the width of the cavity, as well as the composition of the material. Wenzel and Cassie-Baxter models are two major theoretical frameworks that describe how transitions in liquid behavior are depending on surface roughness because wettability is influenced by roughness. The fluid totally drenches the substrate in the Wenzel model. When air is trapped between the cavities in the Cassie-Baxter model, it can cause partial liquid-air and partial solid-liquid interfaces to coexist. [27–29]. The schematic diagram showing both the aforementioned states is shown in Figure 9.1.

9.2 SYNTHESIS OF GREEN CUTTING FLUID (GCF)

Suvin et al. 2020 [14] performed experiments where coconut oil-based Green Cutting Fluid (GCF) was developed using a concoction of non-toxic emulsifiers and sustainable additives to reach the capabilities of commercially available fluids. Experiments were carried out by various researchers to test the efficiencies of various bio-oils. Sunflower oil, coconut oil, and mustard oil have all demonstrated their ability to work with MO in a variety of applications. The inherent tribological properties of coconut oil set it apart from other vegetable oils, including a high flash point, low iodine number, strong oxidation stability, and less free fatty acids. Perera et al. [30] tested two straight oils (soybean and white coconut oil) as well as a MO-based soluble oil for machining. The average coefficient of friction for unrefined and virgin coconut oil samples was marginally lower than for refined coconut oil samples, according to the findings. The difference in the coefficient of friction is caused by the presence of free fatty acids in virgin and unrefined coconut oil. It has been discovered that this environmentally friendly metalworking fluid possesses material properties comparable to commercial metalworking fluid while posing no environmental risks. Hence, Suvin et al. based the development of their nontoxic cutting fluid on coconut oil.

Nontoxic emulsifiers such as TN 85 or Polysorbate 85 (EF-1), TN 80 or Polysorbate 80 (EF-2), and TEA or Triethanolamine (EF-3) were used to create a coconut oil emulsion, as well as additives Azadirachta Indica Oil (A-1), Cymbopogon

```
┌─────────────────────────────────────────────────────┐
│              Emulsifiers (E1 + E2 + E3)               │
└─────────────────────────────────────────────────────┘
                            ⬇
┌─────────────────────────────────────────────────────┐
│                 Azadirachta indica oil                │
└─────────────────────────────────────────────────────┘
                            ⬇
┌─────────────────────────────────────────────────────┐
│                Cymbopogon citratus oil                │
└─────────────────────────────────────────────────────┘
                            ⬇
┌─────────────────────────────────────────────────────┐
│                Centella asiatica additive             │
└─────────────────────────────────────────────────────┘
                            ⬇
┌─────────────────────────────────────────────────────┐
│                 Jaggery syrup (diluted)               │
└─────────────────────────────────────────────────────┘
                            ⬇
┌─────────────────────────────────────────────────────┐
│                      Coconut oil                      │
└─────────────────────────────────────────────────────┘
                            ⬇
┌─────────────────────────────────────────────────────┐
│                   Green cutting fluid                 │
└─────────────────────────────────────────────────────┘
```

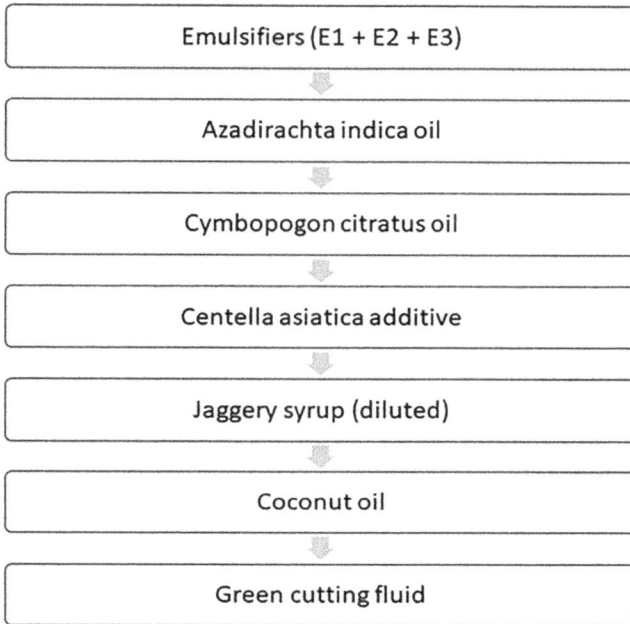

FIGURE 9.2 Flowchart representing formulation of GCF.

Citratus oil (A-2), Ocimum Tenuiflorum oil (A-3), and jaggery syrup (A-4) In the specified concentrated form of cutting oil, EF-1, EF-2, and EF-3 account for 40% of the total weight, while A-1, A-2, A-3, and A-4 account for 20%, and coconut oil accounts for 40%. Because the dilution ratio for this formulation is 1:20, 20 liters of cutting fluid may be made from 1 liter of green cutting oil concentrate. Figure 9.2 shows the schematic formulation of the GCF. CCF is most commonly found on the market. It is mineral-oil based and is routinely used in the metalworking sector [31].

9.2.1 Emulsion Stability

The zeta potential, oven test results, and particle size were utilized to gauge the emulsion stability of green cutting fluid. Zeta potential and particle size were measured using the 90Plus Particle Size Analyzer (Brookhaven Instruments Corporation). Using an oven test in accordance with ASTM D3707-89, the thermal stability was evaluated by gauging how well a 100 ml sample of an emulsion separated into oil and water after being heated at 85 C for 48 hours in a thermostatically controlled oven. The sample's free water, oil, and emulsion content were measured after incubation. To measure viscosity, a Brookfield DV-E viscometer was employed. Using a pH meter, the emulsion's pH was determined.

Deionized water and the cutting oil (coconut oil with emulsifiers and additives combination) were mixed in a 20:1 ratio to make the emulsion. With water, the emulsifiers TN 85, TN 80, and TEA were able to create a stable emulsion. When tested under the same conditions, the findings obtained from DLS equipment by

TABLE 9.1

Characteristic Properties of GCF and CCF

Evaluation Test	Test Sample	
	GCF	CCF
Creamy layer, ml	0	1
Droplet size (Pz), nm	220.1	190
Zeta Potential (mV)	47.46	35.63
Viscosity at 45°C and 100°C (mPa.s)	95.94 and 10.90	21.25 and 3.81
Oven Tests (70–80°C)	Stable	Stable
pH	7.5	9
Odor	Lemon	Mineral oil

measuring particle size and the Zeta potential suggest that GCF was stable and competent to commercially produced samples.

The approximate average particle size of GCF was determined to be 220.1 nm, vs 200 nm for CCF, while the zeta potential of GCF was 47.46 mV, versus 42.23 mV for CCF. The visual appearance of the emulsions was used to measure their storage durability; GCF exhibits the least separation layer after 45 days, whereas CCF developed a creamy layer with an oily base phase. The thermal stability of the emulsion sample was assessed using an oven test according to ASTM D3707-89, which measured the separation of the emulsion sample into oil and water after heating at 85°C for 48 hours in a thermostatically controlled oven [31]. Table 9.1 shows the characteristic properties of the cutting fluids considered.

9.2.2 TESTS PERFORMED

9.2.2.1 Fish Toxicity Test

The toxicity of aquatic organisms was investigated using a fish toxicity test. The test was carried out on two-month-old zebrafish. The fish were given a one-week acclimation period. Initially, the test was performed at a concentration of 100 mg/L. The test was repeated at greater doses (220, 484, 1064.8, and 2342.56 mg/L) to check if the fish were able to survive in the sample conditions. At 24, 48, 72, and 96 hours, fish deaths were recorded, and the concentrations that killed 50% of the fish (LC50) were calculated. Concentrations within 50 (LC50) of more than 1000 mg/L are regarded as largely non-toxic to aquatic systems. The tests were conducted in a controlled environment (27°C, 7 pH) according to the OECD-203 standard.

The acute fish toxicity of coconut oil, emulsifiers (EF-1, EF-2, and EF-3), various green additives (A-1, A-2, A-3, and A-4), GCF, and CCF was investigated. Coconut oil's LC50 value was calculated to be >2342.56 mg/L, while its EF value was assessed to be >1064 mg/L. The LC50 of the GCF with green additions is 1064 mg/L. The LC50 value of CCF is 100 mg/L, and its toxicity is ten times that of GCF. CCF has an LC50 value of less than 100 mg/L and is 10 times more hazardous than GCF.

With 100 mg/L CCF used in a zebra fish survival study, the fish perished within 24 hours, however, when 1064 mg/L GCF was used, there was no toxicity. This finding shows that even at 10 times greater concentrations of GCF, there is no evidence of aquatic toxicity. According to OECD 203 test procedures, GCF succeeds CCF by being nontoxic at LC50 > 1000 mg/L. Even at lower quantities, fish were unable to live in the system using commercial samples; this might be due to toxins present in the cutting fluid. The rise in the LC50 value might be due to the fact that the chemicals that make up the final composition of GCF are non-toxic and environmentally friendly. The toxicity of GCF and CCF was evaluated using fish toxicity tests (OECD 203 standards), and it was discovered that GCF was superior to CCF by being nontoxic at LC50 > 1000 mg/L. [31].

9.2.2.2   BOD and COD Tests

To evaluate the biodegradable capability of newly formulated cutting fluid (GCF) with commercial sample (CCF), biochemical oxygen demand (BOD) and chemical oxygen demand (COD) tests were performed. The amount of dissolved oxygen absorbed by aerobic bacteria in 5 days at 20°C is measured by BOD. The quantity of oxygen was determined using the IS 3025 titration technique. In a COD digester held at 150°C for 2 hours, COD is a measure of dissolved oxygen required for the chemical breakdown of organic molecules. The BOD to COD ratio was used to assess the biodegradability of cutting fluid samples (GCF and CCF). The values of the BOD, COD as well as the ratio, BOD/COD is tabulated in Table 9.2. Non-biodegradable if the BOD to COD ratio is less than 0.3, slow biodegradable if it is between 0.3 and 0.5, and biodegradable if it is greater than 0.3. The GCF had BOD to COD ratios of 1.05, respectively, indicating that it is highly biodegradable, whereas the four commercial fluids examined had non-biodegradable ratios. As a result, the use of non-biodegradable CCF might pose harm to the environment during disposal [31].

9.2.2.3   Corrosion Testing

Corrosion was discovered in the chips, as well as a stain mark on the filter paper. Weight reduction in chips is a measurement of quantity, whereas stain mark observation in filter paper is an assessment of quality with the naked eye. GCF had the least amount of corrosion. Whereas CCF showed evident corrosion on the filter paper as well as in the chips as corrosion products.

The highest corrosion was seen in Deionized water. GCF has the lowest amount of corrosion, as observed by chips and stain marks on filter paper. This finding might be related to the cutting fluid's anti-corrosion properties. TEA served as an emulsifier

TABLE 9.2
BOD/COD for the Considered Cutting Fluids

Sample	BOD (mg/L)	COD (mg/L)	BOD/COD
GCF	42	40	1.05
CCF 5	54	360	0.15

and corrosion inhibitor in GCF. The kind of corrosion inhibitor employed, as well as the combined impact of the base oil, emulsifiers, and additives utilized in the formulation, are all factors that contribute to the anti-corrosion characteristic. While corrosion was obvious in CCF, it was also visible in the chips (as corrosion products).

Weight loss outcomes are documented. A scale of 1 to 10 was used by ASTM D4627 to measure the rate of corrosion, with 1 denoting non-corrosive properties, 5 denoting moderate corrosion, and 10 denoting the most aggressive corrosion. The tests reveal that GCF with grade 3 has superior anticorrosive characteristics when compared to grade 4 of the majority of CCF samples studied. The weight loss research finds that the GCF is more corrosion-resistant than the CCF after a 24-hour incubation period. The outcomes showed that the corrosion rate for GCF is the same as for CCF. On filter paper and chips, all other CCFs and the control showed corrosion stains, while GCF and CCF1 had the least corrosion [31].

9.2.2.4 Drilling Experiments

Drilling experiments were carried out using an HMT FN2 machine. The force and torque were measured using a drill dynamometer from Magnum Instruments. For drilling operations, torque and thrust force were computed. The mild steel (EN8) block was utilized.

For the different combinations of controlled parameters, drilling was done with a 10 mm diameter, high-speed steel drill bit for drilling holes of 30 mm depth. It was determined whether cutting fluid performed better in terms of reduced force and torque by switching from petroleum to bio-based cutting fluid for each set of studies. The torque value fell as the spindle speed was raised, whereas the torque value grew while the feed rate was increased. As more material is removed, increasing the feed rate will raise the surface roughness rating, and vice versa. Both CCF and GCF yielded force and torque values that were comparable and near enough. The push force dropped as the spindle speed rose, whereas the thrust force grew as the feed rate increased. For a sustainable manufacturing process, a cutting fluid should be examined to verify its toxicity and biodegradability levels. These findings corroborate observations and findings from a tribological test done using a tribometer to assess the produced friction during milling. The emulsion's ability to render a low friction layer on the surface is determined by the emulsion's composition as well as the tribological system's speed, load, and temperature properties. Current cutting fluid standards primarily consider machining performance and ignore ingredient selection and biodegradability testing, which are critical for assessing a product's overall quality and long-term viability. Because the force and torque values measured were comparable, it is noted that GCF qualities were as efficient as CCF. As a result, GCF might be a viable alternative to MO-based cutting fluid in the metalworking sector [31].

9.3 METHODOLOGY OF DETERMINING EFFICIENCY THROUGH WETTABILITY STUDIES

In order to determine the efficiency of the GCF, its wettability is tested on three different materials, Aluminum AA5052, Titanium Alloy Ti6Al4V, and EN31 Steel.

In addition to being extensively utilized in machining processes, these materials cover a wide range of different hardness levels. Five samples of each of the materials were obtained of 30 mm × 30 mm × 15 mm dimensions and were initially mirror-polished to remove pre-existing topographies, if any, that might have occurred during machining operations. Silicon Carbide emery sheets with grits of 80, 120, 400, 800, and 1200 were employed to generate new surface topographies. For each of the materials Aluminum AA5052, Titanium alloy Ti6Al4V, and EN31 Steel, the substrates thus generated with defined roughness will be referred to as 80G, 120G, 400G, 800G, and 1200G, respectively. Only unidirectional (UD) roughness was induced by the polishing procedures used. This was accomplished by polishing the samples in unidirection only with emery sheets of the desired grit size. The surface roughness is the major parameter that is being changed. In this way, a total of 15 samples were made, five for each material, in ascending order of grit size.

The obtained surface topographies were analyzed using a non-contact optical profilometer WYKO NT1100. The optical profilometer is a measurement equipment that creates a profile of a reflecting surface using a laser beam reflected from it. The instrument has been configured to allow for stability of the measured profile in the face of laser intensity variations and the reflectivity of the item being examined. Profilometry can be used to determine the extent of the damage. A quantitative evaluation of the wear that has occurred in the time period between two observations is obtained by comparing 3-D maps of the surface profile. With a noncontact optical profilometer, it is possible to make measurements rapidly and without causing damage to the surface. The substrate is placed under the laser beam and is focused in such a way that interference fringes are observed on the screen of the machine. Interference fringes develop in the superposition domain of two diffracted beams. The substrate is then scanned and surface roughness parameters like Ra, Rsk, Del a, Rq, etc were acquired from the profilometry readings and there were utilized to correlate them with the contact angle readings so that a better understanding of the influence of these roughness parameters on wettability can be obtained. Figure 9.3 shows the profilometer images of two samples, EN31-80G and EN31-1200G.

In order to test the wettability of sessile drops, a contact angle goniometer (Dataphysics OCA 30/6) was used. A contact angle goniometer is a device that accurately measures the angle at which a droplet makes contact with a surface. This is a good, indirect way to detect surface wetting. The contact angle of a droplet

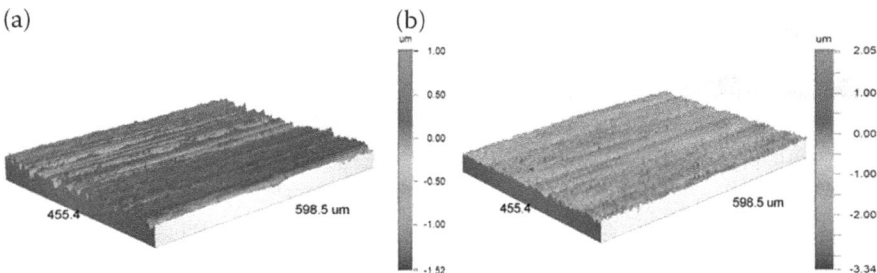

FIGURE 9.3   Profilometer image of (a) EN31-80G (b) EN31-1200G.

becomes 90° as it begins to spread out over a surface. When the droplet is flat, or has a contact angle of 0°, complete wetting occurs. Contact angle goniometers may also be used to produce optical tensiometry measurements by suspending a droplet from the tip of a needle and comparing droplet models to the image obtained by the instrument. This equipment is often used to explore the characteristics of solid/liquid and liquid/liquid interfaces, such as wettability, surface/interfacial tension, surface-free energy, etc. To test the contact angle measurements on any material surface, contact angle readings of water on the corresponding polished surface are used as a standardized reference. Figure 9.4(a-c) shows the liquid bead formation of water and GCF 10% on the polished materials. Figure 9.4(d) shows the corresponding values of Water and GCF 10% on various substrates. With the trials, GCF in a 10% concentration is employed as a lubricant and tested on samples of EN31 steel, titanium alloy Ti6Al4V, and aluminum alloy AA5052. The samples were placed horizontally on the goniometer sample holder. Prior to the procedure of each test, a trial was conducted by dropping a test drop on the substrate to ensure its flatness. Experiments were conducted by dripping 10 µl of the GCF onto the surface of the substrate and permitting it to settle before taking the measurements of contact angles. The optimal amount of lubricant required to create a drop that landed on the surface with apt consistency was observed to be 10µl. The contact angle is acquired when the drop falls from the syringe and drops on the surface. Understanding the lubricant's wettability response on the various substrates under consideration is made easier with the help of these steady-state contact angle data. A thorough analysis of the findings is performed. The procedure is repeated changing the concentration of GCF to 5%. To determine the efficiency of wettability of the GCF, contact angle readings

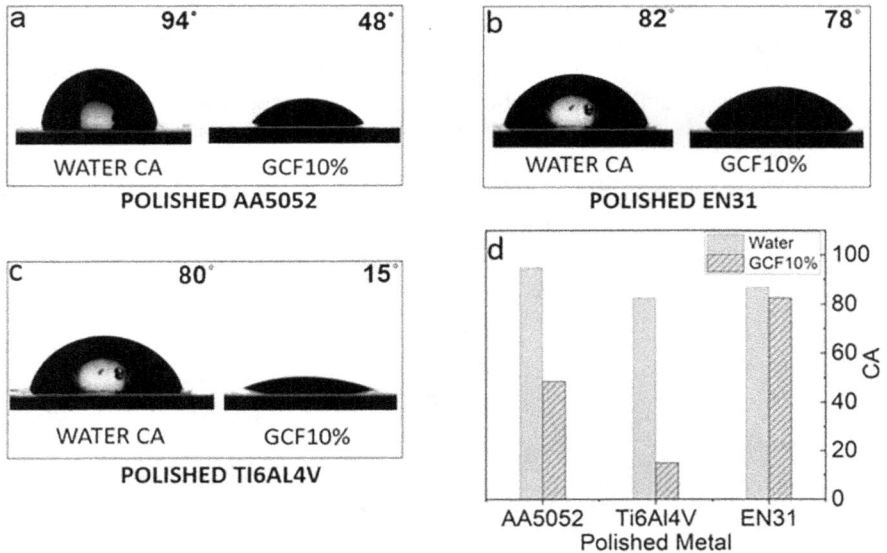

FIGURE 9.4 Water and GCF10% contact angle on (a) Polished AA5052 (b) Polished Ti6Al4V (c) Polished EN31 (d) Polished metal surfaces.

are measured for both the above-mentioned concentrations and compared with CCF contact angle readings of the corresponding concentrations.

9.4 RESULTS AND DISCUSSION

9.4.1 SURFACE TOPOGRAPHY ANALYSIS

Figure 9.3 shows the optical profilometry images of material EN31 Steel of grit sizes 80G and 1200G with unidirectional roughness. It has been noted that the UD method creates continuous grooves along the surface that function as tiny lubricant reservoirs. After metallographic polishing of the samples from each of the materials with different emery ranging from 80 to 1200 grit, the corresponding roughness value Ra as obtained from the optical profilometry study is shown in Figure 9.5c. The average roughness was measured and seen to be in the range of ~Ra 2150 nm (2.1 μm) to 780 nm for Al, ~Ra 625 nm to 163 nm for Ti, and ~Ra 366 nm to 54 nm in steel. It is found that the average roughness values vary indirectly from the hardness of the materials considered. Thus, the Ra values are in the order AA5052>Ti6Al4V>EN31 (Table 9.2).

To aid ease of visualization of the grooves and their arrangement on the surface, bearing area curves were plotted for all sizes of grit and are shown in Figure 9.5 (a and b). The proportion of surface area that can come into contact with a counter-surface at any given height from a predefined reference is represented by the Bearing Area Curve (BAC) [32]. The surface roughness of an item is described by the BAC. Drawing lines parallel to the datum and measuring the proportion of the line that lies within the profile can be used to get the curve from a profile trace.

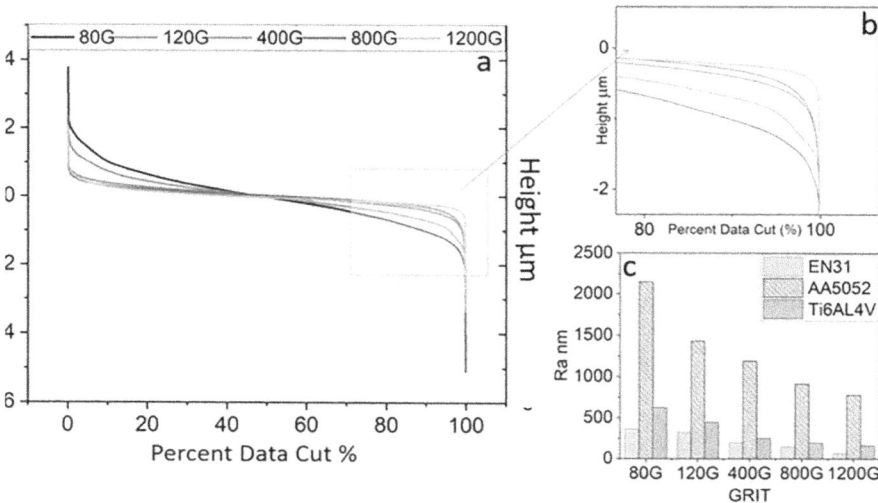

FIGURE 9.5   (a) Bearing area curve (b) Enlarged portion of the bearing curve (c) Average roughness value obtained by corresponding emery grit sheets on roughened surfaces of AA5052, Ti6Al4V, EN31.

The peak height and groove depth rose with increasing roughness, as can be seen by the bearing area curve, and the transition was less abrupt for higher roughness as compared to lower roughness.

From the bearing area curve, the values of MR1 and MR2 can be obtained which aid the precise understanding of the percentage of grooves, that is the valleys and the peaks on the surface. MR1 is the upper limit of the Core Roughness Profile and MR2 is the lower limit of the Core Roughness Profile. These values are tabulated in Table 9.3. Other surface roughness parameters like Rpk - Reduced Peak Height, Rvk - Reduced valley Depth, Rq - root mean square (RMS) roughness, Sm - mean peak spacing, Del a - mean slope and texture density, obtained from the profilometer are also tabulated in Table 9.4. Rpk denotes a surface with many high peaks, producing a tiny initial contact area and, as a result, high contact stress regions when the surface is touched. From the MR2 point to the 100% point, Rvk is the height on the Y-axis of a triangle having the same area as the BAC curve. RMS Roughness, Rq, is the root mean square average of the profile heights over the evaluation length. The mean value of the spacing between profile anomalies within the evaluation length, Sm, is the mean value of the spacing between profile

TABLE. 9.3

Roughness Parameters of (a) AA5052 (b) EN31 (c) Ti6AL4V

AA5052	Ra(um)	Rq(um)	Delax(mrad)	Texture Density(/mm)		Smx(um)	Lam a(m)
				X Cross	Y Cross		
80	2.1	2.8	249.141	44.38	48.89	16.001	18.151
120	1.4	1.98	123.089	31.43	51.59	15.501	18.092
400	1.1	1.59	167.579	54	61.66	13.052	14.613
800	0.9	1.22	165.614	66.39	70.3	14.838	16.848
1200	0.7	1	142.089	59.83	61.96	13.84	15.167
Ti6Al4V	Ra(nm)	Rq(nm)	Delax(mrad)	Texture Density		Smx(um)	Lam a(m)
				X Cross	Y Cross		
80	624.14	789.44	38.04	19.96	69.31	15.209	12.648
120	446.04	574.51	34.672	24.6	92.04	14.952	14.253
400	253.63	337.61	156.24	205.06	232.55	6.949	7.524
800	198.14	283.64	149.424	256.44	260.53	6.142	6.802
1200	163.77	219.36	37.033	65.57	169.15	15.327	12.37
EN31	Ra(nm)	Rq(nm)	Delax(mrad)	Texture density		Smx(um)	Lam a(m)
				X Cross	Y Cross		
80	366.22	478.44	140.565	135.67	204.11	7.72	8.404
120	330.16	426.5	71.948	74.76	162.41	9.606	9.809
400	200.53	286.93	22.854	35.52	129.14	18.667	18.32
800	148.73	188.31	11.342	20.87	153.04	13.222	14.068
1200	69.34	106.07	20.68	102.56	266.6	8.041	9.832

TABLE. 9.4

Bearing Area Parameters of (a) AA5052 (b) EN31 (c) Ti6AL4V

AA5052-Grit	RK nm	Rpk nm	Rvk nm	MR1	MR2	V1 nm	V2 nm
80	6382	3715	2434	0.13	0.91	259	106
120	4000	1922	2770	0.12	0.86	11	185
400	3475	1521	2255	0.10	0.87	81	144
800	2656	1194	1919	0.10	0.88	65	114
1200	2290	740	1394	0.07	0.85	27	104
Ti6Al4V-Grit	**RK nm**	**Rpk nm**	**Rvk nm**	**MR1**	**MR2**	**V1 nm**	**V2 nm**
80	196	933	631	0.10	0.89	51.25	32.66
120	1372	694	606	0.11	0.89	39.83	30.46
400	7808	303	509	0.08	0.88	13.62	28.63
800	5476	280	511	0.10	0.87	14.66	33.01
1200	443	364	228	0.16	0.92	29.56	8.58
EN31-Grit	**RK nm**	**Rpk nm**	**Rvk nm**	**MR1**	**MR2**	**V1 nm**	**V2 nm**
80	1145	385	689	0.08	0.88	16.27	39.32
120	1053	399	579	0.09	0.89	18.58	29.63
400	560	373	510	0.09	0.88	18.61	28.54
800	484	222	191	0.07	0.89	8.52	10.15
1200	198	191	107	0.12	0.91	11.84	4.81

irregularities. The average Absolute Slope, Del-a, is the arithmetic average of the absolute value of the rate of change of the profile height calculated of the evaluation length. Texture Density represents the number of peaks per unit area. A large number indicates more points of contact with other objects.

When the grit size was increased (from 80G to 1200G), it was seen that the surface roughness metrics Rpk, Rvk, Rq, and Del-a followed a declining trend, although texture density increased.

9.4.2 INFLUENCE OF GCF ON WETTABILITY

Figure 9.6 depicts the image of a droplet post-settling, from which the contact angle measurements were obtained. GCF is used in two concentrations, 5% and 10% to comprehend the influence of concentration on the wettability of each of the three different materials. It is observed that the contact angle varies with grit size as well as the composition of the substrate. For this reason, it can be speculated that wettability is also dependent on the composition of the material. When the contact angle is examined, comparing it to the variation in surface roughness, it seems to be that the contact angle decreases with both an increase and a decrease in surface roughness. This is a striking observation in terms of surface lubrication, as the contact angle is a measure of lubricant wettability.

It is observed that in 5% concentration, for both Al and Ti, there is a sudden drop in wettability at 400G whereas in steel 400G, wettability is at its peak. With 10%

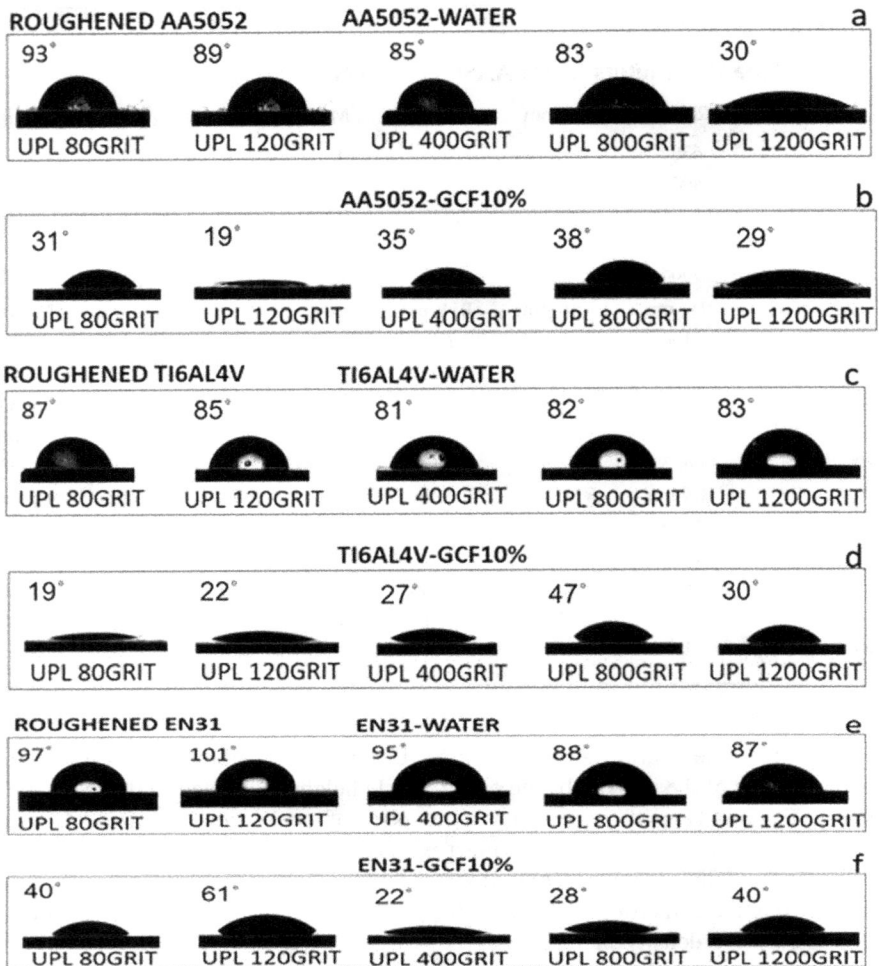

FIGURE 9.6 Contact angle of (a) water on roughened AA5052 (b) GCF10% on roughened AA5052 (c) water on roughened Ti6AL4V (d) GCF10% on roughened Ti6AL4V (e) water on roughened EN31 (f) GCF10% on roughened EN31.

concentration, Al and Steel portray the sudden dip in contact angles at 400G and peak contact angles for Steel with the same grit size.

These alterations in contact angle are the result of the previously described wetting transitions within the mixed wetting regime to near Wenzel or near Cassie-Baxter states. The wetting behavior is largely Mixed Wetting or Composite Wetting because metallographically roughened surfaces contain a variety of roughness frequencies. Between the Wenzel and Cassie-Baxter states is the mixed wetting state. The transition in the 400G of all the materials is thought to be approaching the Cassie-Baxter state if wettability is low and Wenzel state if wettability is at its highest. However, it is also worth noting that the contact angle might display values in a very narrow range for different Ra values produced by the same grit emery. These findings point to the necessity for more research into the surface profile characteristics in order to effectively correlate these discrepancies.

9.4.3   Effect of Concentration of GCF

With the interest of obtaining the most favorable wetting properties, the green cutting fluid is tested in two different concentrations. The 5% and 10% concentrations of the lubricant are formulated by mixing 5 and 10 parts of the GCF, respectively with 100 parts of distilled water. The obtained cutting fluids were tested with each of the samples and the contact angle measurements were obtained. Using the contact angle measurements, the wettability was calculated for both of the concentrations. The results were analyzed and compared. The graphs – as shown in Figure 9.7 – clearly indicate higher contact angle readings for 5% concentration in all three materials. The inverse of the contact angle readings gives the wettability of the lubricant. The values of the percentage difference in each grit size are shown in Figure 9.8. Mean percentage differences of 36%, 55%, and 48% were observed for AA5052, Ti6Al4V, and EN31, respectively. Hence, it can be concluded from the graphs that in all the cases, a 10% concentration of the GCF shows higher wettability when compared to 5% of the GCF.

9.4.4   Comparative Assessment of GCF and CCF

When evaluating the effectiveness of the working mechanism of cutting fluids, one of the key deciding criteria is how cutting fluids interact with metal surfaces. The CCF is a widely employed, well-known cutting fluid having the ability to be used in a number of metallurgical operations. This cutting fluid's replacement should have wetting properties on par with or better than the CCF. The contact angle readings of the CCF 5%, GCF 5%, CCF 10%, and GCF 10% were obtained from the sessile-drop experiments done with the goniometer. Figure 9.9 depicts the comparative analysis of contact angle measurements between the two considered cutting fluids on each of the materials considered. Evaluating the performance of both the cutting fluids in 5% concentration, on all three materials, it is observed that in this particular concentration, CCF shows greater wettability. However, when comparing the lubricants'

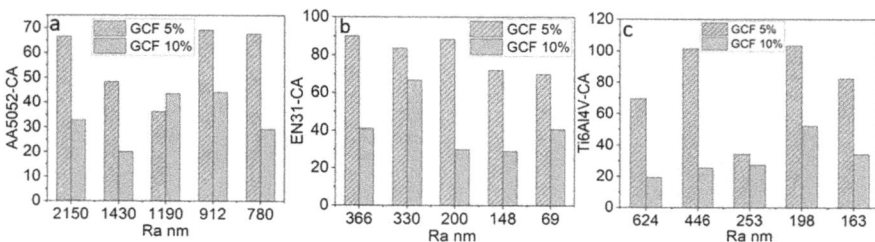

FIGURE 9.7   Comparison between GCF5% and GCF10% on roughened (a) AA5052 (b) EN31 (c) Ti6AL4V.

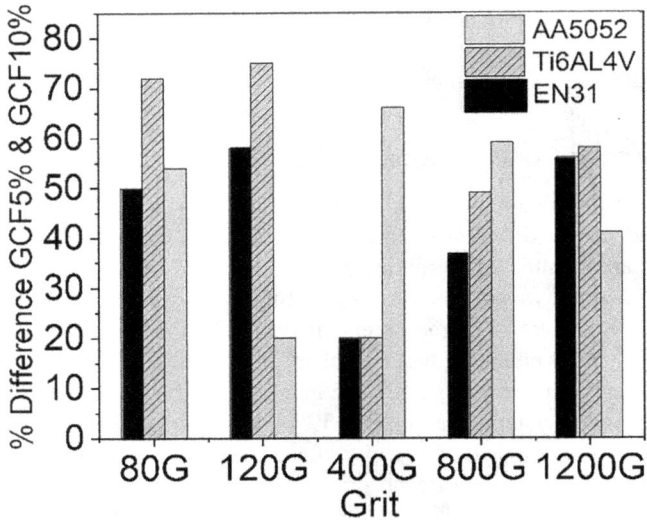

FIGURE 9.8   Percentage difference between GCF 5% and GCF 10% concentration.

10% concentrations, the wettability significantly outperforms that of the CCF at the same percentage. For 80G, 120G, 400G, 800G, and 1200G, respectively, there is a wettability increase of 76%, 53%, 40%, 30%, and 16% for Ti. Al and Steel exhibit a comparable level of wettability improvement at a 10% Green Cutting fluid concentration (Figure 9.10). The commonly observed order of wettability is GCF 5% < CCF

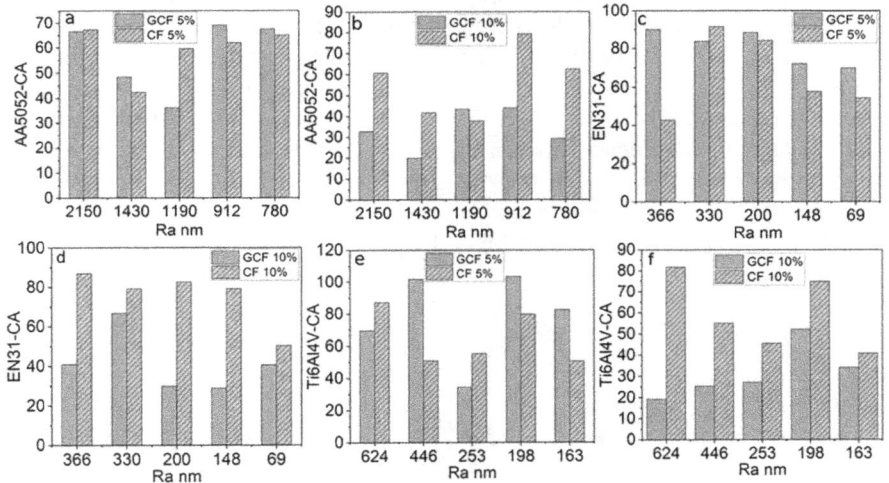

FIGURE 9.9   Comparison between (a) GCF5% vs CF5% on roughened AA5052 (b) GCF10% vs CF10% on roughened AA5052 (c) GCF5% vs CF5% on roughened EN31 (d) GCF10% vs CF10% on roughened EN31 (e) GCF5% vs CF5% on roughened Ti6AL4V (f) GCF10% vs CF10% on roughened Ti6AL4V.

(a) (b)

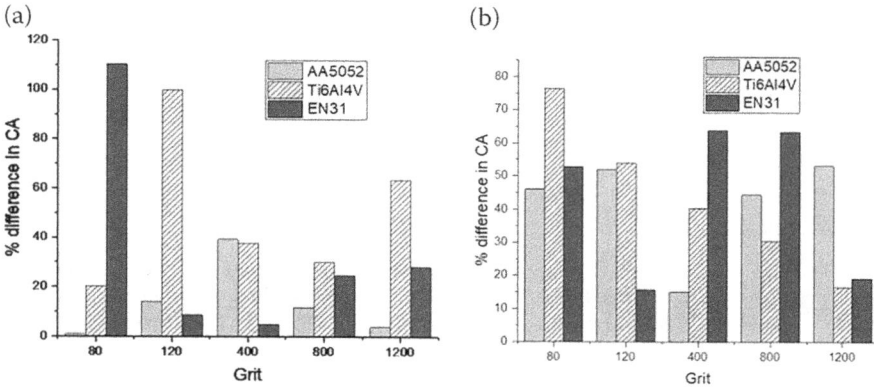

FIGURE 9.10 Percentage difference in wettability between a) GCF 5% and CCF5% b) GCF 10% and CCF 10%.

5% < CCF 10% < GCF 10%. Because of the high wettability characteristics of GCF 10%, it can be concluded that in addition to providing environmental advantages, GCF may be used as a workable substitute for CCF because of its higher wettability and potential to facilitate machining operations.

9.5 CONCLUSIONS

Experiments on wettability were conducted on unidirectional-roughness-induced surfaces of Aluminum AA5052, Titanium alloy Ti6Al4V, and EN31 Steel to examine the efficiency of indigenously developed green cutting fluid. From the research, the following findings may be drawn.

1. GCF in 10% concentration portrayed a mean percentage increase of 36%, 55%, and 48% wetting for Al, Ti, and Steel respectively when compared to GCF in 5% concentration. Hence GCF 10% is a more optimal concentration to be considered for machining operations.
2. Wettability results show that surface roughness is capable of increasing as well as decreasing the wettability of cutting fluid. Peaks in contact angle are observed at 120G in Steel and 800G in Ti and Al. This suggests the influence of roughness dependence on wettability has different trends in different roughness ranges and also depends on the composition of the substrate material.
3. It was discovered that surface roughness was the main factor affecting the contact angle created by the GCF drop for the unidirectionally textured surfaces on each of the three samples used for the investigation. Thus, a correlation exists between measurements of contact angle and surface roughness.
4. The results of the comparative evaluation between the CCF and GCF showed that the GCF had a significant percentage increase in wetness.
5. In addition to the coconut oil base of the GCF being eco-friendly, all the required components used for the effective formulation of the GCF were

also eco-friendly. Hence the GCF is nontoxic, biodegradable, and environmentally friendly as it is created from renewable resources.

6. The findings provide strong, clear proof that the GCF, with its benefits for the environment and improved wetting properties at 10% concentration, is a practical, long-term replacement for MO-based cutting fluids.

REFERENCES

[1] R. Somashekaraiah, P. S. Suvin, D. P. Gnanadhas, S. V. Kailas, and D. Chakravortty, "Eco-Friendly, Nontoxic Cutting Fluid for Sustainable Manufacturing and Machining Processes," *Tribology Online*, vol. 11, no. 5, pp. 556–567, Sep. 2016, doi: 10.2474/TROL.11.556

[2] V. Edachery and S. v Kailas, "Influence of surface texture directionality and roughness on wettability, sliding angle, contact angle hysteresis, and lubricant entrapment capability," *Tribology International*, vol. 158, p. 106932, 2021, doi: 10.1016/j.triboint.2021.106932

[3] E. O. Aluyor and M. Ori-jesu, "Biodegradation of mineral oils – A review," *African Journal of Biotechnology*, vol. 8, no. 6, pp. 915–920, Sep. 2010, doi: 10.4314/ajb.v8i6.59986

[4] M. R. Sankar, J. Ramkumar, and S. Aravindan, "Machining of Metal Matrix Composites with Minimum Quantity Cutting Fluid and Flood Cooling," *Advanced Materials Research*, vol. 299–300, pp. 1052–1055, 2011, doi: 10.4028/WWW.SCIENTIFIC.NET/AMR.299-300.1052

[5] A. Kumar, K. K. Gajrani, P. S. Suvin, S. K. Vasu, and M. R. Sankar, "End Milling of Biodegradable Magnesium Alloy (AZ 31) with Minimum Quantity Eco-friendly Cutting Fluids".

[6] S. Pervaiz, S. Kannan, and H. A. Kishawy, "An extensive review of the water consumption and cutting fluid-based sustainability concerns in the metal cutting sector," *Journal of Cleaner Production*, vol. 197, pp. 134–153, Oct. 2018, doi: 10.1016/J.JCLEPRO.2018.06.190

[7] E. Kuram, B. Ozcelik, and E. Demirbas, "Environmentally Friendly Machining: Vegetable Based Cutting Fluids," pp. 23–47, 2013, doi: 10.1007/978-3-642-33792-5_2

[8] J. J. Eppert, K. L. Gunter, and J. W. Sutherland, "Development of cutting fluid classification system using cluster analysis," *Tribology Transactions*, vol. 44, no. 3, pp. 375–382, Jan. 2001, doi: 10.1080/10402000108982470

[9] E. Kuram, B. Ozcelik, E. Demirbas, E. Şik, and I. N. Tansel, "Evaluation of new vegetable-based cutting fluids on thrust force and surface roughness in drilling of AISI 304 using Taguchi method," *Materials and Manufacturing Processes*, vol. 26, no. 9, pp. 1136–1146, Sep. 2011, doi: 10.1080/10426914.2010.536933

[10] U. S. Dixit, D. K. Sarma, and J. P. Davim, "Machining with minimal cutting fluid," *SpringerBriefs in Applied Sciences and Technology*, no. 9781461423072, pp. 9–17, 2012, doi: 10.1007/978-1-4614-2308-9_2

[11] H. S. Abdalla, W. Baines, G. McIntyre, and C. Slade, "Development of novel sustainable neat-oil metal working fluids for stainless steel and titanium alloy machining. Part 1. Formulation development," *International Journal of Advanced Manufacturing Technology*, vol. 34, no. 1–2, pp. 21–33, Aug. 2007, doi: 10.1007/S00170-006-0585-4

[12] I. S. Foulds, "Cutting fluids," *Kanerva's Occupational Dermatology, Second Edition*, vol. 2, pp. 715–725, Jan. 2012, doi: 10.1007/978-3-642-02035-3_64

[13] S. Debnath, M. M. Reddy, and Q. S. Yi, "Environmental friendly cutting fluids and cooling techniques in machining: a review," *Journal of Cleaner Production*, vol. 83, pp. 33–47, Nov. 2014, doi: 10.1016/J.JCLEPRO.2014.07.071

[14] P. S. Suvin, P. Gupta, J. H. Horng, and S. v. Kailas, "Evaluation of a comprehensive nontoxic, biodegradable and sustainable cutting fluid developed from coconut oil:" *https://doi.org/10.1177/1350650120975518*, vol. 235, no. 9, pp. 1842–1850, Nov. 2020, doi: 10.1177/1350650120975518

[15] L. Wang, X. Tian, Q. Lu, and Y. Li, "Material removal characteristics of 20CrMnTi steel in single grit cutting," *Materials and Manufacturing Processes*, vol. 32, no. 13, pp. 1528–1536, Oct. 2017, doi: 10.1080/10426914.2017.1279298

[16] Y. Mao and J. Wang, "Is green manufacturing expensive? Empirical evidence from China," *International Journal of Production Research*, vol. 57, no. 23, pp. 7235–7247, Dec. 2019, doi: 10.1080/00207543.2018.1480842

[17] R. Katna, K. Singh, N. Agrawal, and S. Jain, "Green manufacturing—performance of a biodegradable cutting fluid," 10.1080/10426914.2017.1328119, vol. 32, no. 13, pp. 1522–1527, Oct. 2017, doi: 10.1080/10426914.2017.1328119

[18] L. K. Toke and S. D. Kalpande, "Critical success factors of green manufacturing for achieving sustainability in Indian context," *International Journal of Sustainable Engineering*, vol. 12, no. 6, pp. 415–422, Nov. 2019, doi: 10.1080/19397038.2019.1660731

[19] R. Katna, M. Suhaib, and N. Agrawal, "Nonedible vegetable oil-based cutting fluids for machining processes–a review," *Materials and Manufacturing Processes*, vol. 35, no. 1, pp. 1–32, Jan. 2020, doi: 10.1080/10426914.2019.1697446

[20] H. S. Abdalla and S. Patel, "The performance and oxidation stability of sustainable metalworking fluid derived from vegetable extracts," *Proceedings of the Institution of Mechanical Engineers, Part B: Journal of Engineering Manufacture*, vol. 220, no. 12, pp. 2027–2040, 2006, doi: 10.1243/09544054JEM357

[21] A. Kalantarian, R. David, and A. W. Neumann, "Methodology for high accuracy contact angle measurement," *Langmuir*, vol. 25, no. 24, pp. 14146–14154, Dec. 2009, doi: 10.1021/LA902016J

[22] R. E. Johnson and R. H. Dettre, "Contact angle hysteresis. III. Study of an idealized heterogeneous surface," *Journal of Physical Chemistry*, vol. 68, no. 7, pp. 1744–1750, 1964, doi: 10.1021/J100789A012

[23] G. Whyman, E. Bormashenko, and T. Stein, "The rigorous derivation of Young, Cassie–Baxter and Wenzel equations and the analysis of the contact angle hysteresis phenomenon," *Chemical Physics Letters*, vol. 450, no. 4–6, pp. 355–359, Jan. 2008, doi: 10.1016/J.CPLETT.2007.11.033

[24] M. S. Bell and A. Borhan, "A Volume-Corrected Wenzel Model," *ACS Omega*, vol. 5, no. 15, pp. 8875–8884, Apr. 2020, doi: 10.1021/ACSOMEGA.0C00495

[25] H. Yildirim Erbil and C. ElifCansoy, "Range of applicability of the wenzel and cassie-baxter equations for superhydrophobic surfaces," *Langmuir*, vol. 25, no. 24, pp. 14135–14145, Dec. 2009, doi: 10.1021/LA902098A/SUPPL_FILE/LA902 098A_SI_001.PDF

[26] D. Murakami, H. Jinnai, and A. Takahara, "Wetting transition from the cassie-baxter state to the wenzel state on textured polymer surfaces," *Langmuir*, vol. 30, no. 8, pp. 2061–2067, Mar. 2014, doi: 10.1021/LA4049067/SUPPL_FILE/LA404 9067_SI_004.AVI

[27] V. Hisler et al., "Model Experimental Study of Scale Invariant Wetting Behaviors in Cassie–Baxter and Wenzel Regimes," *Langmuir*, vol. 30, no. 31, pp. 9378–9383, Aug. 2014, doi: 10.1021/LA501225M

[28] J. Jopp, H. Grüll, and R. Yerushalmi-Rozen, "Wetting behavior of water droplets on hydrophobic microtextures of comparable size," *Langmuir*, vol. 20, no. 23, pp. 10015–10019, Nov. 2004, doi: 10.1021/LA0497651

[29] B. He, J. Lee, and N. A. Patankar, "Contact angle hysteresis on rough hydrophobic surfaces," *Colloids and Surfaces A: Physicochemical and Engineering Aspects*, vol. 248, no. 1–3, pp. 101–104, Nov. 2004, doi: 10.1016/j.colsurfa.2004.09.006

[30] G. I. P. Perera, H. M. C. M. Herath, I. M. S. J. Perera, and M. G. H. M. M. P. Medagoda, "Investigation on white coconut oil to use as a metal working fluid during turning:" *http://dx.doi.org/10.1177/0954405414525610*, vol. 229, no. 1, pp. 38–44, Apr. 2014, doi: 10.1177/0954405414525610

[31] Suvin, P. S. "Synthesis and testing of eco-friendly, non-toxic cutting fluid emulsions," *PhD diss.*, 2021.

[32] V. Edachery *et al.*, "Enhancing tribological properties of Inconel X-750 superalloy through surface topography modification by shot blasting," *Materials Performance and Characterization*, vol. 10, no. 2, Feb. 2021, doi: 10.1520/MPC20200172

10 Ergonomic Design Analysis of Analogue Micrometer Screw Gauge

Sangeeta Pandit and Neel P Padia
Department of Design, Indian Institute of Information
Technology Design and Manufacturing Jabalpur,
Madhya Pradesh, India

CONTENTS

10.1 INTRODUCTION

Throughout the history of humans, we have always been involved in utilizing resources around us to make objects. From basic necessities such as tools, utensils, clothing and shelter to much more advanced objects such as electronics, medical devices, weapons etc. are all examples of this attribute of human being called creativity. Humans

DOI: 10.1201/9781003242291-10

understood that in order to realise any idea from imagination into reality, resources need to be seen from a different perspective.

Usually, a creation consumes various forms of resources in different stages of its development for example it may require things like raw materials, energy, human resources, chemicals and many more. Consumer appliances, fashion, machinery, medicine, and scientific and specialized equipment are a few of the fields that were explored with different perspectives, such as new materials, processes and science.

But with the passing of time, we have started to realize the ill effects of mass creation on our surroundings, i.e., our planet. Pollution, excessive waste disposal, harm to wild and sea life are few to name. Thus, mindful actions based on the model of sustainability and circularity are in much need to be included in the process of creation.

Manufacturing sector here has a wide scope for implementation of sustainability on both micro as well as macro level. On the micro level one of the areas in the manufacturing sector which usually deals with the identification of 'what has to be rejected as waste' and 'what has to be accepted' comes under the common practice of quality control. Quality control is the process which ensures quality of output as per set standards/criteria using measurement activities. Increase in the efficiency of the quality control process and equipment used in the process could help to achieve some sustainability goals by saving valuable time, reducing wastage and money over a longer period.

More information on metrology and quality control process (part of metrology) is described below.

10.1.1 METROLOGY

The term metrology means 'the scientific study of measurements' [1]. It is derived from two Greek words, 'metron, i.e., measurement' and 'logos that is the study'. It links all human activities worldwide by establishing a commonly understandable unit [2]. The study of measurement is an essential requirement in science and technology, most notably in engineering and manufacturing. The initial concept of metrology is said to have formed during the Egyptian period way back in 2900 B.C. [3,4] when a permanent standard was established using black granite in the form of a cubit. Builders used these cubits as the standard of building. These cubits were made according to the length and width of the pharaoh's forearm and hand, respectively. While a political motivation during the French Revolution aimed to harmonize units throughout the nation laid the foundation of modern metrology for the world. In March 1791, a length standard based on a natural source was proposed for the first time, which we now recognize as 'meter' [5]. Metrology creates broader impacts on a number of sectors such as industry, manufacturing, economics, energy, the environment, health, and most importantly, consumer confidence [6]. The impacts of metrology on the economic exchange are the most apparent societal impacts. Therefore government, concerned agencies and responsible treaties worldwide verify and test standards against a recognized quality system in specialized calibration laboratories to work out the reasonable and exact exchange between nations [7].

So, metrology can be summarised as the definition, realisation and application of acceptable units of measurement through international collaboration [2]. Metrology undoubtedly plays a significant role when it comes to supporting innovation. Metrology encourages the optimization of existing systems through challenges and innovations. Without an agreed measurement system, it would not be possible to trade accurately and fairly, firms will find it challenging to innovate and compete, and government regulation will be ineffective [8]. Each manufacturer uses a set of calibrated measurement instruments for inspections. Calibrations are essential for achieving precision. It ensures that parts of the instrument fits together and works the way it is intended. After the calibration, the manufacturing machines are set, and the quality assurance (Q.A.) and quality control (Q.C.) are established. In a business, quality assurance ensures that quality requirements will be comprehended, while quality control checks that the requirements have been achieved during the making [9]. Thus Q.C. is strictly defined as 'checking' to ensure that requirements for a manufactured part have been met. In order to meet the requirements of quality assurance, technically advanced and up-to-date instruments are essential to reduce the risks. Even a minute error in measurement can ruin an entire project.

10.1.2 QUALITY CONTROL

Q.C. is a procedure or set of procedures intended to review and ensure the quality of all factors involved in production or services performed according to the quality standards decided. ISO 9000 defines it as 'A part of quality management focused on fulfilling quality requirements' [10]. Q.C. is often usably interchanged with Q.A. Both are two aspects of quality management, in which the activities can be interrelated, but there are distinct differences between the two. Q.A. is focused on forming a clear vision through planning, documenting and confirming a set of guidelines. It eventually helps the organization to eliminate testing problems and assure quality. On the other hand, Q.C. is simply conducting tests to verify the quality of the output. While Q.A. deals with the specified requirements to be met by a product or service, Q.C. is one of the parts that helps achieve it through inspection [11]. Inspection remains a significant component of quality control, where a physical product is examined visually or methodically. Product quality inspectors are provided with established standards that the produced part must satisfy to qualify before reaching customers. Based on data generated during the inspection, companies prepare plans to improve service and quality so that no poor service or product reaches customers. Finally, the Q.C. process must be continuous to guarantee the prompt discovery of repetitive issues and the restorative endeavors, whenever required, to produce agreeable outcomes [12].

10.1.3 ERROR IN MEASUREMENT

An error can be defined as an incorrect or inaccurate action [13]; it is also synonymous with a mistake in some practices. While an 'error' is a deviation from the correct value, a 'mistake' is an error caused by a fault. The fault can be due to any reason like carelessness or misinterpretation [14]. Statistics does not consider an error as a mistake but rather a difference between a resulting value and the accepted [15]. However, a degree of uncertainty is always there in the resulting value for any measurement. All measurements are inaccurate, and the error caused is simply the difference between the indicated and actual values [16]. In a measuring system, each component can act as a potential source of errors and contribute to overall measurement error.

Errors in a measurement system are broadly classified into two types, viz., random error and systematic error [17]. Random error is consistently present in an estimation. They occur randomly, and the specific source for the same is usually indeterminable. However, a possible cause for such error can be inherently unpredictable variances in the readings or possible friction between parts of a measuring instrument or the observer's interpretation of the reading with an analogue scale [18]. For example, measuring someone's height is affected by minor posture changes. Opposite to random error, systematic errors, which are also referred to as avoidable errors, are predictable and typically constant or proportional to the true value. Systematic errors can be the cause of various reasons like interference of the environment in measurement, improper calibration of measuring instruments, or imperfect observation methods [19]. For example, an improperly calibrated thermometer may give accurate readings within a specific temperature range but become inaccurate at higher or lower temperatures. It is further classified into instrument error, environment error and observation error. Instrument errors during measurement may occur due to imperfections in the instrument's design, defects in the manufacture of adjustment of an instrument or improper instrument calibration [20]. Due to various reasons, errors are caused in components combined and lead to measurement errors. For example, errors in temperature measurement, due to poor thermal contact between the substance whose temperature is to be found and the thermometer.

Observational errors, which is also termed as parallax error sometimes, occur when a line of vision of the observer and the pointer above the reading scale does not match or when the observer fails to correctly identify, measure accurately, or interpret some aspect of the phenomena that are being observed [18]. Observation error mainly occurs with analogue instruments. For example, a good example of parallax can be seen in the dashboard of a vehicle that uses a needle-style (analogue) speedometer. When viewed directly in front, the speed may show let's say 60, but when viewed from the passenger seat, the needle may show a slightly different speed due to the angle of viewing. Environmental errors happen due to the outside situation of the measuring instruments. It has different causes such as temperature, humidity, magnetic or electrostatic field, vibrational interferences, wind and improper lighting. For example, a Geiger counter in a lab may show greater radiation when a cell phone is near than what the actual radiation is, as cell phones radiate RF waves, which disrupts the reading.

10.1.4   Measuring Instrument

Nowadays, organizations have various quality tools (for example, measuring equipment, report books, testers, etc.) to assist them in continuous improvement and update of their procedures. These tools transform the measurement output into important information that can help in decision making [21]. This way organizations keep the quality system up-to-date and maintain certain business advantages over their competitors [22]. Different measuring instruments provide their outputs differently, some of them can directly indicate the measured value while others give a number relating the measurement with an established standard. Measuring instruments may use separate material measures for, e.g., a weight balance that uses a standard mass and compares it to an unknown mass, or they may contain markings to recreate the unit like a measuring cup or a graduated ruler. Different measuring instruments can range from simple objects such as a stopwatch and a weighing scale to electron microscopes and particle accelerators. However, for the scope of this chapter, we focus on one such measuring instrument, analogue micrometers. It is highly preferred over high-tech digital micrometers in small and medium scale industries of India due to economic constraints.

10.1.4.1   Analogue Micrometer Screw Gauge

A micrometer screw gauge, sometimes known as a micrometer, is used to measure very small objects' depth, length, and thickness. These are often used in engineering labs and machining industries. The word micrometer is derived from Greek words 'micros = small' and 'metron = measure'. The first micrometric screw designed was utilized in a telescope to gauge angular distances between stars and the relative sizes of celestial objects. Later 'Henry Maudslay' fabricated a bench micrometer somewhere around the early 19th century that was jocularly nicknamed 'the Lord Chancellor' among his staff since it would give the final judgement on measurement in the company's work [23]. Micrometers are more precise than alternative metrology tools such as a Vernier Caliper. It can measure distances with very high precision using a calibrated internal lead screw for gradual adjustment. Micrometers are usually callipers type instruments. Their opposing ends are joined by a frame, except in some designs. The object to be measured is placed between a very accurately machined screw (the spindle) and the anvil. The spindle moves by turning the thimble or ratchet knob until the object to be measured is lightly touched by both spindle and anvil.

10.1.4.2   Other Available Micrometers in the Market

Other micrometers for special purposes are also available in the market as – Blade micrometer for diameters inside narrow grooves measurement; inside micrometer – calliper type for small internal diameter and groove width measurement; spline micrometers for splined shaft diameter measurement; tube micrometer for pipe thickness measurement; point micrometers for root diameter measurement; screw thread micrometer for effective thread diameter measurement; disc type outside micrometer for root tangent measurement on spur gears and helical gears; ball tooth thickness micrometer for measurement of gear over-pin diameter and V-anvil micrometer for measurement of 3- or 5-flute cutting tools.

Modern time micrometer screw gauge comes with high-tech electronic sensors, digital display and IoT-based data storage and transmission capabilities for more connected industry management. Some upcoming digital micrometers are bench based with smart digital linear encoders. Such micrometer features a comprehensive software program for controlling test speed, auto zero, auto sample detect, selectable opening distance, selectable measurement range, selectable dwell time, built-in auto sample feed option, statistics, multiple languages and calibration records. The latest version of the micrometer is the laser scan micrometers. The laser scan micrometer uses basic principles of optical distortion. The laser built into a laser scan micrometer sends out a rotating optical beam and has both the originating rotating beam and a receiver located opposite the laser source. The measurement is calculated when a certain area is blocked in the laser path, the object being measured interrupts the light from the laser, making a shadow that represents the object's size based on the length of time of the obstruction. These micrometers are very expensive and have a niche market in comparison to regular micrometers.

10.1.5   ERGONOMIC ASSESSMENT OF A MICROMETER

To analyse a micrometer from an ergonomic perspective, we need to understand its working principle and its construction. A micrometer screw gauge, as its name indicates, utilizes a screw to transform small linear distances [24] into a sizable angular motion that is adequately large enough to read from a calibrated scale. Micrometer derives its measuring accuracy from the accurately machined thread form, which is at the core of its design.

The operating principle behind this is – the axial movement of the screw inside the body of the micrometer screw gauge directly and precisely correlates with the amount of rotation of the screw through a constant known as 'lead'. A screw's lead is the distance pushed ahead axially in one complete turn of the screw (i.e., 360°). This constant is properly calibrated (adjusted), thus allowing micrometers to amplify small distances into large readable formats. Much more important precision is acquired using a differential screw in certain micrometers [24,25].

A micrometer is composed of:

Frame - Micrometers frame is a C-shaped body that maintains a constant relation between the anvil and barrel. It is heavy and has a high thermal mass and an insulating pad to prevent significant heating up by the handgrip; it also helps prevent any distortion in the instrument.
Anvil - It holds the sample/object at rest with the help of a spindle.
Spindle - The cylindrical component moves towards the anvil when the thimble operates.
Screw - As discussed in the working principle, a threaded screw is the core of any micrometer. It is connected to a spindle on one end and is operated by a thimble connected to it on the other end.
Thimble - It is operated using the index finger and the thumb. It is primarily used to move the spindle.
Ratchet - is an arrangement that limits applied pressure by slipping at calibrated torque and is positioned at the end of the barrel close to the thimble.

Since micrometers are used for fine precision measurement it is manufactured using ductile cast iron, carbon steel, stainless steel, high carbon chrome steel and carbide-tipped measuring faces [26]. Micrometers are very sophisticated instruments and in order to ensure its optimum working capabilities companies manufacturing it have set a series of guidelines that need to be followed while using it. For the reference let's consider - 'Mitutoyo Inc.', leading maker of measuring instruments from Japan. Mitutoyo Inc. give out some notes that need to be taken care off while using micrometer, [27] and the notes are as follows:

- Carefully checking type, range, accuracy and other specs as per the application.
- Maintaining temperature for both instrument and workpiece before measurement (controlled environment).
- Both the micrometer and its standard should be left at the same area for at least several hours prior to adjusting the start point.
- Maintaining proper eye level and viewing angle to avoid parallax error.
- Cleaning measuring faces (zero setting) as well as the measuring instrument before measurement for proper maintenance.
- Using the constant-force device (ratchet) accurately so estimations are performed with the right estimating force.
- Using Stand to reduce the influence of temperature change due to body heat and increase stability.
- Being cautious; not to drop or knock the micrometer on anything.
- Not to turn the micrometer thimble utilizing excessive force.
- Taking necessary precautions to prevent any corrosion of anvil such as applying oil on a regular basis.
- Storing the micrometer in a proper case and when storing, always leaving a gap of 0.1 to 1 mm between the measuring faces. [27]

10.1.6 ROLE OF ERGONOMICS IN MAN MACHINE INTERACTION

When we talk about the interaction of man and machine, the discipline of human factors and ergonomics inevitably comes into the picture. The poor ergonomic design may decrease productivity with slower work and more errors. An increase in injuries to the wrist, forearm, and shoulders may occur, which increases over time with the use. If hand, wrist or seating posture deviates from neutral posture, it increases the likelihood of injuries [28]. Similarly, the number of repetitions, application of higher force, and stress level increase the injury level. **These injuries are generally termed musculoskeletal disorders (MSD)** (Figure 10.1) [29]. In general, the risk factors for hand or wrist MSDs are due to higher number of repetitions, excessive force exertion, awkward hand or wrist postures, and hand arm vibration. Common musculoskeletal disorders include – Carpal Tunnel Syndrome, Tendonitis, Tension Neck Syndrome, Muscle/Tendon strain and Trigger Finger/ Thumb. **'Carpal Tunnel Syndrome'** (Figure 10.1) is the most commonly studied MSD among all [30] wherein the subject experiences tingling, numbness, weakness, or muscle atrophy in hand and fingers as the median nerve at the wrist gets compressed, swelling the synovium; which takes up space of the carpal tunnel [30].

Research has proved that individuals can work effectively infrequently and are affected by factors such as posture, duration of task and repetitions [28]. Design of an instrument or tool dectates how it could be handled by a person (Grip/hold) and how it can be operated. For the proper handling of any instrument, each grip can be classified by its need for precision or power.

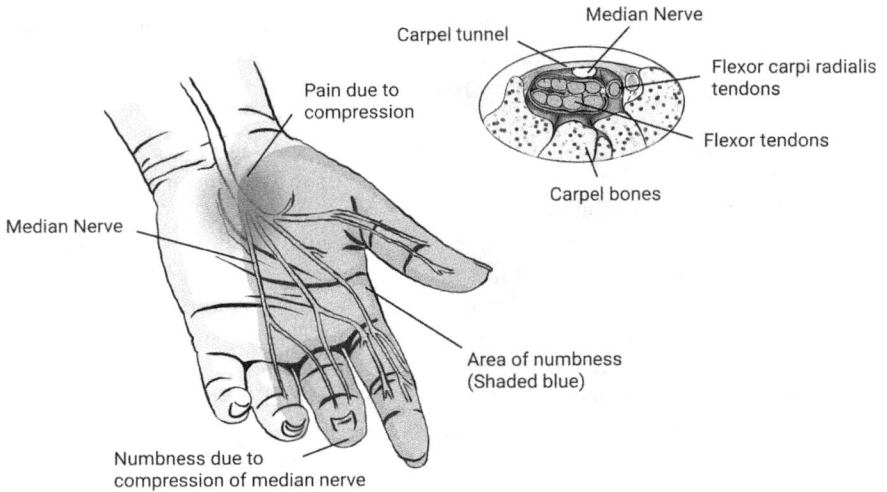

FIGURE 10.1 Nerves and carpal tunnel in human hand and the effect of carpal tunnel syndrome in hands.

In the power grip, there is an unbending connection between the object and the hand, which implies that all movements of the object have to be evoked by the arm [31]. For precision handling, the hand is able to perform intrinsic movements on the object without having to move the arm. Forceful gripping and grasping have been associated with the risk of MSDs such as carpal tunnel syndrome [32]. It is estimated that joint and tendon loading can be five times greater for a pinch type grip than for a power grip [33]. Prolonged exposure to high-force high-precision grips can also lead to localised muscle fatigue and discomfort, which negatively affects a person's productivity [34].

10.2   PROBLEM AREA

India is making much effort to level up its manufacturing capabilities under the 'Make in India' campaign [35]. Indian industries have started to gear up to compete against the global market with more and more focus on homegrown research and development work [36]. Along with multinational companies, many medium and small-scale enterprises and local industries are putting their focus on quality production output [37]. Measuring Instruments like Micrometer are designed and meant to be used with a specific set of guidelines; thus, the challenges offered to the measured output are factors of 'an' Industry environment infrastructure. Problem associated with Micrometer are discussed separately below.

10.2.1   ISSUE WITH MICROMETER GRIP

In order to use a micrometer, the user has to hold it in his/her dominant hand (mostly right hand), placing the C-shaped frame against his/her palm and little or ring finger around the frame (Figure 10.2). The index finger and thumb are principally used to

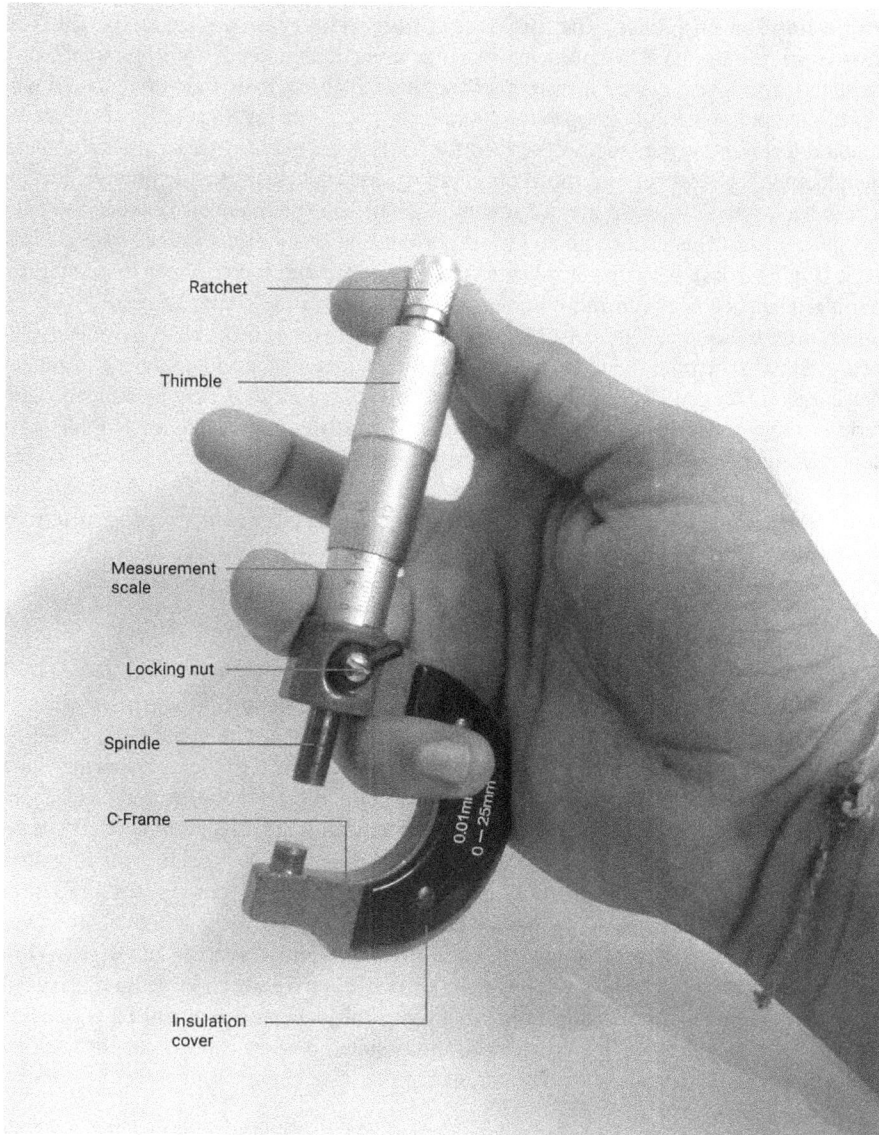

FIGURE 10.2 Hand grasp posture commonly involved with the use of an analogue micrometer screw gauge.

operate the thimble [38]. The other or non-dominant hand of the observer can be used to support the instrument and hold the object to be measured. The operator has to support the weight of the micrometer in his/her palm, which usually ranges from 200 grams to 500 grams (different companies) for 0–25 mm measuring range micrometer [39] and use the thumb and index finger to operate the thimble or ratchet (if finger extension permits). **Although there is no set standard as to how micrometers should be held in a**

single-handed use case, the most common grip type observed is the one shown in Figure 10.2. Micrometer manufacturers recommend using a ratchet or a constant force device during measurement so that the object to be measured is held with appropriate pressure, and the operator could record accurate readings [27]. There are no standard techniques or methods described for holding the micrometer by the micrometer manufacturers. However, in most observed cases, the existing micrometers design impels the operator to grasp the instrument in a very unusual manner: between the palm and little finger. This creates an awkward posture that is not ergonomically suitable for work (Figure 10.2). The operator has to use his thumb and index finger to operate the thimble. Distance between the thimble (and ratchet) and the frame increases with the object's size measured (Figure 10.2). The operator has to focus on holding the object to be measured with the other hand during the measurement. Constant force device or ratchet that is recommended to be used while taking measurement thus becomes inaccessible during single-handed use of the instrument. Micrometers with ratchet thimble have started to come into the market but are twice as expensive as a base version of micrometer (which are widely used in the Indian market due to affordable prices) [40]. Lastly, the little finger or ring finger used to grip the frame against the palm acts as an obstruction for the object.

10.2.2 ATTENTION AND STABILITY PROBLEM

Micrometers' design has evolved from a bench setup to a handheld portable device [41], but micrometer manufacturers still recommend using a stand to hold the instrument while taking measurements. It leads to more accurate measurement and makes it easier to take measurements, reducing random jiggling effects [42]. Manufacturers also state that changing the support method and orientation of a micrometer after zero setting can influence estimating results. However, micrometer stands are bulky and not suitable for portable usage. These stands are designed to be used as a fixed setup for repetitive measurement. A handheld method is generally considered convenient and often required when the measurand workpiece cannot be brought to the worktable. However, holding the micrometer in hand introduces more potential sources of error. **While holding a micrometer in hand, it becomes difficult to provide a stable base, give full attention to measurement, align the workpiece (object), and maintain a consistent measuring force** (Figure 10.3). In addition, as stated earlier, it becomes difficult for operators to reach the ratchet on the far end of the instrument (Figure 10.3).

10.2.3 THERMAL HEAT TRANSFER ISSUE

Holding a micrometer in hand could lead to more disadvantages, such as causing the frame to expand as the heat from the operator's hand can also transfer to the micrometer frame, leading to thermal expansion and ultimately measurement error [42]. Abbe's principle states that 'maximum accuracy is obtained when the scale and the measurement axes are common'. The reason is that any change in the relative angle of the moving measuring faces of a measuring instrument causes displacement that cannot be read on the instrument's scale and this induces an error termed 'Abbe error' [27,42]. **The Figure below illustrates expansion of the**

FIGURE 10.3 Hand posture and attributes of operator while using the micrometer screw gauge and handling the job (object).

micrometer frame due to heat transfer from hand to frame when the frame is held with a bare hand, resulting in a significant measurement error due to temperature-induced expansion (in microns). Therefore, the contact time should be minimized if the micrometer needs to be held by hand during measurement (limiting within 15 mins only) [27]. An error is also induced if a different operator with a different hand temperature holds the same micrometer kept at a standard room temperature [16]. Manufacturers, however, recommend using a stand, gloves or holding the instrument by heat insulation over frame to reduce such errors.

10.2.4 Visual Ergonomics Issue

Every analogue instrument has chances of wrong visual misinterpretation of reading, known as human error. The term 'human error' is often used in metrology and in reference to a measurement that was 'not intended by the user' [13]. This issue can significantly be resolved with the introduction of electronics and sensor technology in the design of instruments [43], but at the same time, it increases the cost of the equipment. In addition, several other instruments, including micrometers, are required in a Q.C. lab. Therefore, the total cost of such high-tech instruments and their calibration is way higher than any analogue instrument, which could cost a significant amount to the organisation.

A human error can occur due to many reasons, for example - stress, lack of concentration, lack of attention, lousy design, environmental factors, fatigue, multitasking. The definition of human error has very loose boundaries and it may sometimes also interfere with other errors such as instrumental error. A human can unknowingly carry out an error due to any persisting design flaw in the measuring instrument [44]. For

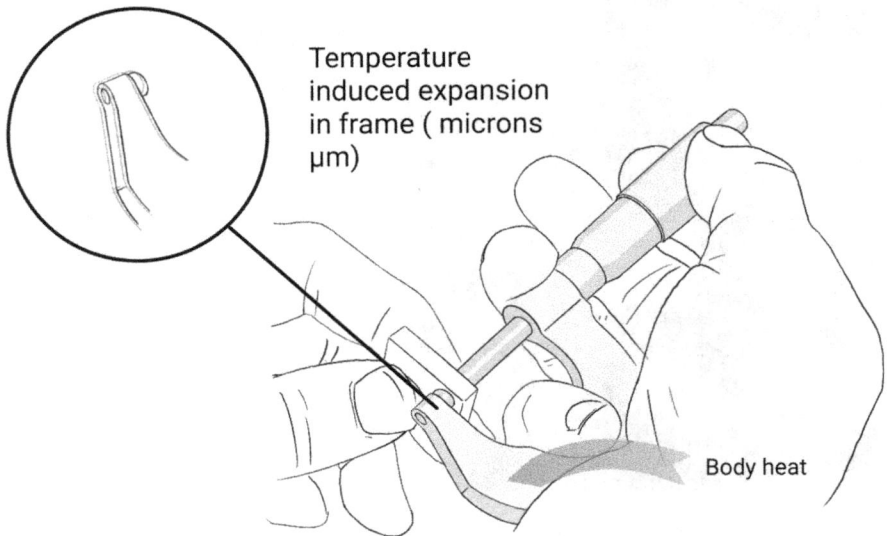

FIGURE 10.4 Illustration representing frame expansion in microns (µm) due to heat transfer from operator's hand to micrometer frame.

example, parallax error occurs when the observer's line of vision is not perpendicular to the scale [45]. One might question the need for the line of vision to be perpendicular to the scale. **The reason behind the parallax error is the faulty design at the front of**

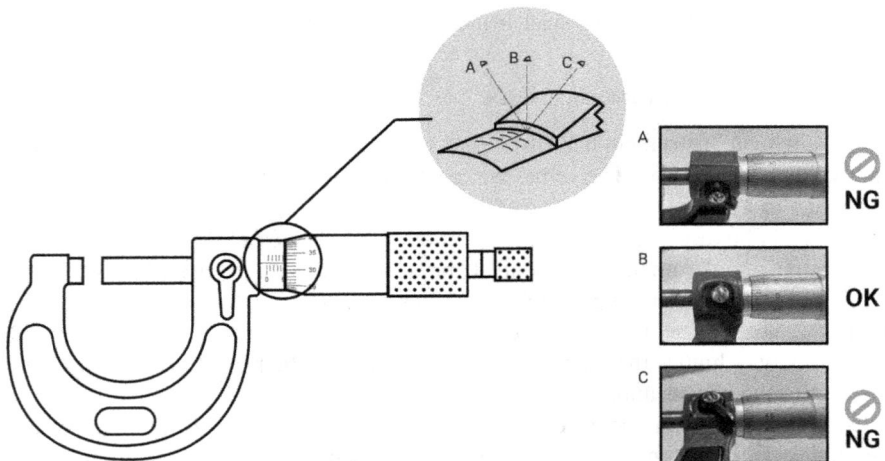

FIGURE 10.5 Visual representation of parallax error issue with Micrometer. If the watching angle is A it is NG (Not good), if it is perpendicular to scale B, then it is OK (Okay) and if it is at angle C it is again NG (Not good).

the thimble, which is slightly raised above the sleeve level (Figure 10.5) [16]. **This causes distortion when viewed from an angle.**

It helps to state that sometimes, the design element or constraints could also lead an observer to conduct an unintended action. If these issues related to visual elements such as scale, identification marks, size, colour, contrast and finish come into the picture, these can be referred to under a common term called visual ergonomic issues [46,47]. In comparison, the rest of the physical aspects leading to any other issue goes under the term physical ergonomics issue (Discussed in section 10.2.1 and 10.2.2).

10.3   TO ERR IS HUMAN

'The idea that a person is at fault when something goes wrong is deeply entrenched in society'. When major accidents occur, official courts of inquiry are set up to assess the blame, which is attributed to 'human error'. 'According to Donald A. Norman, human error usually results from poor design'.

Over time, instruments have evolved into measuring a millionth of a centimeter. Several technological advancements are made on measuring instruments to minimize the chances of human error, but on the flip side, it significantly affects the price and affordability of instruments. Another downside of this approach is that it overlooks the interaction between the human and the instrument (usability aspects), which can potentially cause human error. Although humans cannot always be responsible for human errors, other factors such as poor design and other external factors compel humans to execute errors.

The poor ergonomic design of micrometer is more likely to cause users discomfort leading them to execute human errors than neglection of usage guidelines.

10.4   METHODOLOGY

In order to find the usability issues with the interaction of micrometer, the study was conducted on individuals having experience of using micrometer screw gauge in the Q.C. process. Participants for the experiment were chosen from different companies using non-probability convenience sampling. A total of 20 male participants ranging from 22 yrs to 47 yrs participated in the initial research phase. No specific age limit was set for participants. Their work revolved around inspection activities such as collecting samples from production batches, measuring dimensions and report generation. The average working time of each participant was between 8 to 10 hrs. All of them had a minimum of one year of working experience in quality control during the time of the study.

Quality control inspectors perform the inspection by bringing samples from different batches using inspection devices such as micrometer for checking the dimensional parameters. A Q.C. inspector usually holds a micrometer in one hand and the sample Object to be measured on the other hand. He repeats this activity on different sample objects at different time intervals throughout the day.

This study obtained the data using surveys and interviews through online and offline mediums. The survey questionnaire was prepared based on the design satisfaction questionnaire, the discomfort rating scale [46] and the hand and finger discomfort assessment map [47].

Survey-1 consisted of two sections – section 1 covers questions regarding participants' basic information, work experience, education background, job and industry profile. These questions were necessary to obtain an idea of participants' overall health and familiarity with the field. Section 2 consisted of questions regarding micrometer, skill background and their understanding of usage guidelines. These questions were used to obtain an insight on the type of instruments they are using, participants' acquaintance with formal practices and their current practices (Annexure-10.1).

Participants for the second phase of the survey were selected based on their response to using analogue micrometers. 10 out of such 14 participants agreed upon further participating in the survey. Survey-2 consisted of three sections and was designed more thoroughly with an increased focus towards understanding participants' experience. The first section was the Discomfort rating scale (DRS). It was used to obtain an idea of the overall discomfort felt by users. The second section was a design aspect satisfaction question. It was used to identify physical and visual ergonomic design issues persistent with a micrometer. The third and the last section was the hand and finger discomfort map. It was used to identify sections of the hand experiencing the most discomfort and pain (Annexure-10.2).

Key points extracted from both surveys were used for further qualitative research, i.e., the personal Interview. Each participant was asked a series of open-ended questions (Annexure-10.3), and the response for which was recorded to understand in depth the problems they faced as users.

10.5   RESULT

The research experiment was carried out in three phases – 1st was a basic survey of the participants, 2nd was an in-depth survey regarding problems and 3rd phase was an interview with participants followed by empathy mapping.

Outcome of survey 1

- All the participants surveyed came from different categories of Industry. However, most participants belonged to local medium and small-scale industries (10 and 6, respectively), **while only 4 participants were associated with MNC** (Figure 10.6(b)).
- 15 out of 20 participants were healthy and young, below the 30 years age group. However, 8 out of 20 participants had weaker eyesight and needed to wear corrective lenses during work.
- 4 out of 20 participants were relatively experienced, with more than five years of working experience in quality control and inspection. However, 12 out of 16 remaining participants were relatively new, working in the field only for the last 1–2 years.
- The survey revealed that 14 participants received formal training in Q.C. inspection but were noted not wearing gloves while taking measurements as usually instructed during training. This indicates that their work quality might suffer irregularity due to certain behavioural traits.

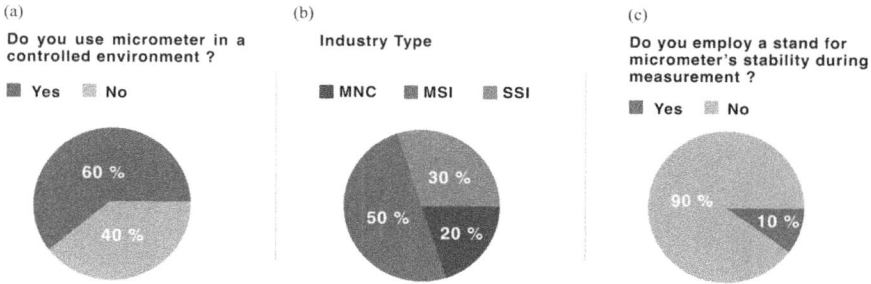

FIGURE 10.6 Response from the survey, 6(a): Percentage of participants using micrometers in a controlled environment. 6(b). Percentage of participants associated with different industry types. 6(c). Percentage of participants using a micrometer stand during the measurement.

- **8 participants expressed that they do not always measure in a controlled environment** (Figure 10.6(a)), which is one of the mandated instructions given during training or by the makers of micrometers. Therefore, irregularity is caused due to external factors.
- 14 participants use only the analogue version of the micrometer, and 6 participants use advanced digital micrometers, while only 2 of them use both (as per availability). Most of the participants using analogue versions belong to medium and small-scale industries.
- **Surprisingly 18 out of 20 participants never employed a holder (stand) for micrometers while taking measurements** (Figure 10.6(c)), while 15 participants admitted that taking micrometers for measurement in different locations is quite often. This reports against the guidelines given by micrometer manufacturers and training sessions – 'that to avoid errors that could arise due to improper stability and hand heat transfer it is advisable to use a micrometer with a stand'.

Outcome of survey 2
Discomfort Rating scale

- **The survey showed that all 20 out of 20 participants found the micrometer screw gauge uncomfortable to a certain degree while being used (refer to** Figure 10.7).

Design aspect satisfaction questions
Physical design aspects

- **Out of the 7 design features, grasp force, material finish and weight of the instrument were evaluated as satisfactory by the majority of the participants. Out of 10 participants, 5 were satisfied with grasp force** (Figure 10.8(a)), 5 with the material finish and 6 with the weight of the micrometer.
- 7 out of 10 participants found grasp span/length as narrow. This indicates

Please mark the level of discomfort felt by you while using micrometer for longer time

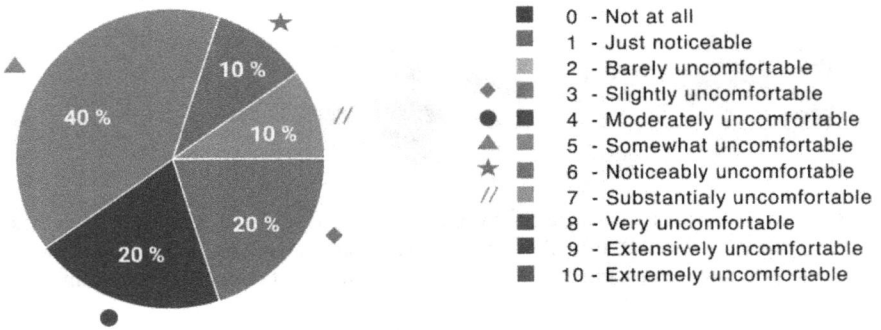

- 0 - Not at all
- 1 - Just noticeable
- 2 - Barely uncomfortable
- 3 - Slightly uncomfortable
- 4 - Moderately uncomfortable
- 5 - Somewhat uncomfortable
- 6 - Noticeably uncomfortable
- 7 - Substantialy uncomfortable
- 8 - Very uncomfortable
- 9 - Extensively uncomfortable
- 10 - Extremely uncomfortable

FIGURE 10.7   The amount of discomfort experienced by participants in percentage.

(a)   (b)

Grasp force requirement -   **Distance of ratchet from frame -**

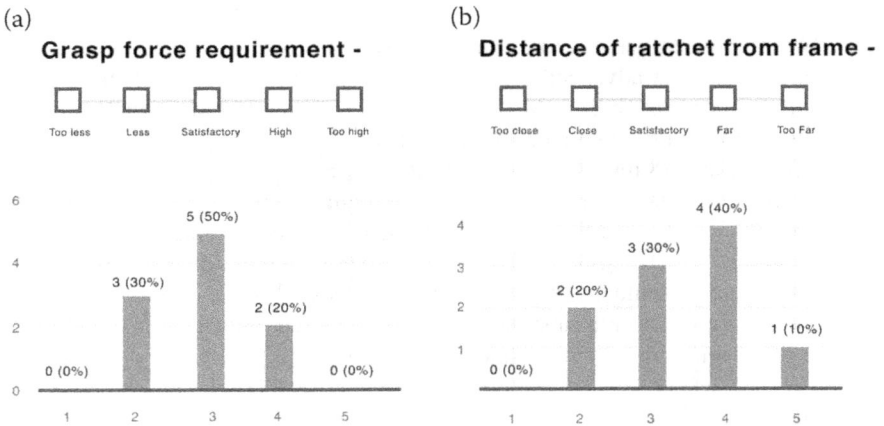

FIGURE 10.8   (a) Participants' opinion regarding Grasp force requirement while holding Micrometer. On the x-axis is the level of satisfaction from 'Too less' to 'Too high' and on the y-axis is the 'Number of participants'. (b) Participants' response regarding the distance between the ratchet and micrometer frame. On the x-axis is the level of satisfaction from 'Too close' to 'Too far' and on the y-axis is the 'Number of participants'.

the users could have been experiencing difficulty in comfortably holding the instrument.

- Participants had mixed reviews about frame orientation: 4 participants found it satisfactory, 4 flat, and the rest of 2 found it steep. This further adds to confirm the difficulty experienced by participants holding micrometer properly.
- **10 participants experienced that the distance between the ratchet and the frame (place to grip the instrument) was much longer/far** (Figure 10.8(b)), making it difficult to operate the constant force device –

ratchet as instructed by makers of micrometers. Moreover, in the case of single-handed use, the user had to undergo an awkward grip posture.

*   6 out of 10 participants felt the frame's shape to be fit for use, and the remaining 4 participants found it to be not fit for use. In addition, a significant number of participants found difficulty holding micrometer screw gauges supporting the fact that the existing micrometer design offers ergonomic challenges to its users.

Visual design aspects

*   7 out of 10 participants were not satisfied with the micrometer's marking scale (division). That means instrument users could have been experiencing eye strain while taking measurements. Only 1 participant felt it satisfactory to read the scale.
*   **Overall, 9 out of 10 participants found micrometer screw gauge to offer low readability** (Figure 10.9(a)).
*   6 out of 10 participants were comfortable with the marking orientation of the micrometer. The remaining 4 participants thought that the marking orientation was incorrect. While using a micrometer in one hand, after securing the workpiece between anvils and spindle using a thimble, participants were needed to turn the instrument in order to read the measurement.
*   **7 out of 10 participants have experienced trouble due to the surface reflectivity of the reading scale** (Figure 10.9(b)). In addition, external

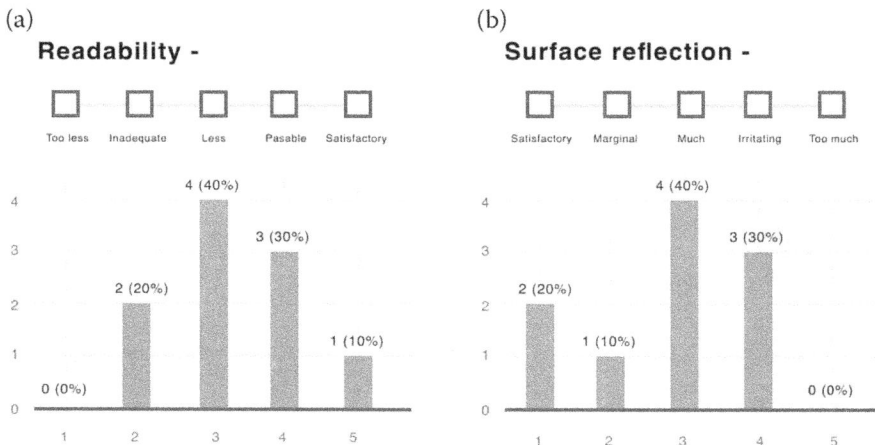

FIGURE 10.9 (a) Participants' opinion regarding readability of the micrometers scale. On the x-axis is the level of satisfaction from 'Too less' to 'Satisfactory' and on the y-axis is the 'Number of participants'. (b) Participants' response regarding surface finish reflection. On the x-axis is the level of satisfaction from 'Satisfactory' to 'Too much' and the y-axis represents the 'Number of participants'.

effects such as atmospheric light and intensity cause unnecessary reflections creating disturbances during measurement.

Discomfort in hand area

- **9 out of 10 participants experienced discomfort in the Hypothenar area (12)** (Figure 10.10).
- 8 out of 10 participants experienced discomfort in the Little finger distal Phalanx (1), Little finger middle and proximal phalanges (2), Thumb distal Phalanx (9), Thumb proximal Phalanx (10), Metacarpal area (11) and Thenar area (13) (Figure 10.10).
- 6 out of 10 participants experienced discomfort in the Ring finger middle and the proximal phalanges (4) (Figure 10.10).
- 4 out of 10 participants experienced discomfort in the Ring finger distal Phalanx (3), Index finger distal Phalanx (7) and Index finger middle and proximal phalanges (8) (Figure 10.10).
- The rest of 2 out of 10 participants experienced discomfort in the Middle finger distal Phalanx (5) and the Middle finger middle and proximal phalanges (6) (Figure 10.10).

Outcome of Interview

Quality control is a job that demands focus and attention from the one who does inspections and measurements since the reputation of any company depends on the quality of its output; companies rely on their crew for its affirmation. During the interview, participants associated with medium and small industries proclaimed that the functional aspects of the organisation are compromised due to limited infrastructure. Due to lean financial settings, companies have to use outmoded equipment, while other similar companies in the competitive environment may use far better

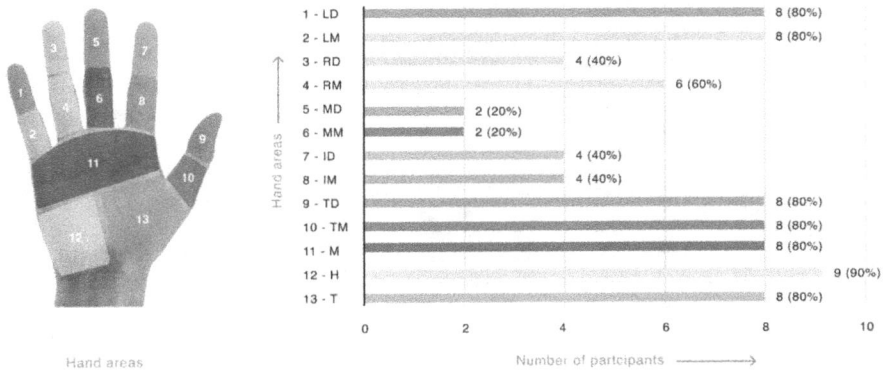

FIGURE 10.10 Discomfort experienced in different areas of hand (13 areas) amongst the sample set (10 participants) in percentage. On the x-axis is the 'Number of participants' and on the y-axis are 'Different areas of Hand'.

equipment for high-quality outputs. Many participants who took part in the research were associated with automobile ancillary machining companies; thus, a '1 inch' analogue micrometer was among the most commonly used instruments for taking measurements as micrometers are principally used for precise measurement of diameter and thickness.

Participants highlighted during the interview that order parts are made in batches and each vendor company offers a certain set of tolerances; thus, the QC department carries out inspection activity multiple times a day. Different companies have different systems; some source samples from batches to a commonplace while others promote observers to inspect the machining. A micrometer is a portable device allowing it to be taken to different places, although it is advised to take measurements at a common location in a controlled environment to eliminate unnecessary fluctuations. Micrometer stands provide stability but is bulky and not portable. Only two participants admitted to using a micrometer stand in the survey, the rest reported unavailability due to limited infrastructure. Some interviewees (participants from small industries) reported that only a few small-scale companies have a dedicated area with a reasonably controlled environment that facilitates inspection activity. To roll out faster production, many companies carry out inspection activities in unfavourable environments, resulting in lower efficiency of error rejection. During the interview, experienced quality inspectors explained that the quality of any final output is a factor of man, machine (instruments), and management. Proper instrumentation overcomes the shortcomings of the other two and helps achieve quality output, which is a deciding factor for a company's reputation.

It is not surprising to hear Q.C. personnel not following guidelines such as wearing gloves, using a micrometer stand for stability, using constant force devices regularly or thermally stabilizing the equipment in different environments before using micrometers for inspections. Even though 70% of participants obtained some form of training, it was discovered that the credibility of training came into question when many participants failed to follow a set of guidelines associated with the proper usage of instruments. As per the statement of Don Norman, these guidelines can be dispensable, having the design of micrometer such that it covers all of these shortcomings.

These elements significantly impact the way a user uses this instrument and the output generated. The influences created by these design elements are reported in participants' quotes during the interview.

"Our hands are sometimes dirty or oily; at that time, the chances of the instrument slipping of hands are high, and the texture on the frame of the micrometer is simply not rough enough to help maintain a firm grip".

Many participants expressed that the micrometer has an unusual design and thus feels uncomfortable after using it for longer durations. Few participants stated having to use balm and massage for pain relief after repetition of the activity. Different people have different hand sizes, and a smaller frame may not suit everyone. A participant expressed –

"I am taller than usual, and my hand size is comparatively larger than many. The 1-inch micrometer that I often use for work feels quite small for my hand. As a result, my little finger becomes obstruction during measurement".

Some of the participants feel that the micrometer frame is not designed correctly. Results of Survey-2 showed the areas of the hand affected in order to hold and grip the micrometer between the little finger and palm. During the interview, it was found that almost all participants experienced pain and discomfort in their hands after work.

"I know it is advisable to do measurements using a ratchet, but in single-handed use, it becomes difficult to reach the ratchet while holding the frame. Holding a micrometer in one hand and also operating it and also at the same time holding the object and concentrating, that is when we tend to make mistakes sometimes while recording measurements".

When participants were asked to use a ratchet, many participants reported it is difficult to use a ratchet in single-handed use. In addition, if the measurement is large (more than 15mm), the position of the thimble and ratchet moves further away with respect to the frame. In case the observer uses both hands to operate the micrometer, turning the ratchet knob becomes easy, but holding the object and micrometer together becomes challenging, thus increasing the chances of executing errors. The stability issues remain persistent if micrometer stands are not employed during measurements.

8 participants in survey-1 reported having weaker eyesight and a need to wear glasses. Such participants expressed during the interview that although good eyesight is one of the job requirements in inspection activities, not everyone using a micrometer has perfect eyesight. Hightech digital micrometers, which reduce reading errors, are available in the market, but it does not solve the problem because digital micrometers are expensive. In addition, its timely calibration increases the cost of the companies more than what an analogue micrometer could.

"I am not young. It's hard for me to read anything written in small size without glasses. I have to look closely to be sure and it causes strain to the eyes".

The environment where the micrometer is used also affects the visual ergonomics factors. Using the micrometer under a bright sunny environment creates reflections over the surface of the micrometer scale (marking), creating glare during reading. In addition, the markings on the micrometer are quite small and thin, which forces users to look closely in order to avoid misreading. Due to these reasons, many participants feel that the readability of micrometers is poor. Many participants expressed difficulty in reading as the marking orientation is horizontal. Hence participants have to turn the instrument every time to note the reading, which could be easily avoided if a micrometer stand is used. Ergonomic design and behavioural aspects together create a lot of effect on measurements taken by micrometers.

10.6 DISCUSSION

The results of the study correspond to some of those in existing research. Research has shown that the efficiency of measurements depends on the 'linkage' between the measuring devices and human operators [48]. However, no specific information is available regarding the micrometer screw gauge. In the past, few attempts have been made to minimise the effect of poor ergonomic design of micrometers. Results obtained from the discomfort rating scale (Figure 10.7) and hand area discomfort map (Figure 10.10) correspond with the finding that the grip force required should get distributed to as large a pressure-bearing area on the finger and palm as possible to increase the work efficiency [49,50]. In the Patent US 2013 009 1720A1 [51], the issue of thermal expansion due to body heat transfer over a long time of use was addressed since it is known that unnecessary error can be generated due to thermal expansion, affecting the measurement efficiency of the micrometer [27]. Dissatisfaction with grip span and frame shape (see section 10.5.) also correlates with the finding that the smaller the point of support, the greater pressure the user must apply to hold devices we regularly use such as a stylus. The greater the pressure, the user experiences more strain and discomfort on their hand [52]. Experience noted by participants over continuous usage of micrometers resonates with the finding that repetitive gripping with high forces and improper postures may significantly increase the risk of work-related musculoskeletal disorders at the upper extremity [46]. Key insights from the interview showed that gloves and a stand to reduce heat transfer are not popular among users. Figure 10.4 shows the degree of expansion with time and the possibility of affecting measurements. Figure 10.8(b) shows low satisfaction with comfortability, due to the large distance between the ratchet and the frame (in case of single-handed usage). Insights from the interview also reported a lack of stability while taking measurements and holding objects together. This observation corresponds with the motive behind patent US4485556 [53], US 8.245,413 B2 [54] & US9482509 [55]; all these designs facilitate one-hand operation and stability. It does not restrict the degree of freedom of the other hand (used to hold objects) so that the operating efficiency can be improved.

Problems similar to lack of readability (Figure 10.9(a)) in micrometers because of the smaller size of marking, were also associated with reading of a 'Mercury Manometer' [56] and 'Haematocrit readings using reflexion baematocritometry' [57]. Interview responses also showed that scale orientation caused users to turn around and look closely to record measurements from the scale. The effect of external elements such as surface reflection also affected readability. Responses from surveys and interviews correspond with studies of the human factors of measuring devices; physiological and psychological human characteristics can significantly influence the way humans use any tool or do work [48].

It was observed that due to ergonomic issues, users have to experience substantial pain, numbness and discomfort in their little finger, ring finger, thumb and palm. Since no previous study has been conducted on the ergonomic analysis of micrometer and risks of grip posture involved with its operation, findings from other awkward grip postures and ergonomics issues in the task such as pencil grasp, stylus operation, dental device operation, kitchen pincers and plier operation are used to compare with the results of this study. Study data indicated that continuous use of micrometers leads users to execute errors due to the risk of musculoskeletal disorder or carpal tunnel syndrome. This study was limited to quantitative and qualitative data but should be further prompted for ergonomics testing and analysis.

10.7 CONCLUSION

Existing literature on the study of human factors and ergonomic analysis of micrometer screw gauges is scarce; only a few patented innovations are related to improving the usability aspect of the micrometers. Several research findings focus on investigating the importance of the 'human factor' in designing various instruments and tools, but study on analogue micrometer is scarce. The present study confirms the instrument's design's lack of ergonomic, behavioural, and environmental factor considerations. Analogue micrometers are preferred over high-tech digital micrometers due to the lack of infrastructure and economic constraints in small and medium-scale industries. However, some instances of error were noted amongst study participants due to uncomfortable feeling experienced over continuous usage of a micrometer. The present study confirms that a detailed ergonomic study in the design of the analogue micrometer is the need of the hour with the application of proper industrial design principles.

REFERENCES

[1] H. Czichos, T. Saito, and L. E. Smith, eds., *Springer Handbook of Metrology and Testing*. 2nd edition. Berlin, Germany: Springer, pp. 3, 2011.
[2] P. Howarth, and F. Redgrave, *Metrology - In Short*. 3rd ed. Albertslund: Schultz Grafisk, p. 9, 2008.
[3] H. Czichos, T. Saito, and L. E. Smith, eds., *Springer Handbook of Metrology and Testing*. 2nd edition. Berlin, Germany: Springer, p. 23, 2011.
[4] "A Brief History Of Metrology". 2021. Bowers Group. Co.Uk. https://www. bowersgroup.co.uk/row/news/brief-history-of-metrology/.
[5] "History of Measurement, French-Metrology." n.d. https://web.archive.org/web/20110425025041/http://www.french-metrology.com/en/history/history-mesurement.asp.
[6] EURAMET-European Association of National Metrology Institutes. n.d. "Metrology for Society's Challenges." Euramet.Org. Accessed June 30 2021. https://www.euramet.org/metrology-for-societys-challenges/?L=0
[7] H. Czichos, T. Saito, and L. E. Smith, eds., *Springer Handbook of Metrology and Testing*. 2nd ed. Berlin, Germany: Springer, p. 30, 2011.
[8] K. Robertson, and J. Swanepoel, "The Economics Of Metrology," *Australian Government or the Department of Industry, Innovation and Science*. 2005. https://www.industry.gov.au/sites/g/files/net3906/f/June%202018/document/pdf/the_economics_of_metrology.pdf
[9] J. M. Juran, and A. Blanton Godfrey, *Juran's Quality Handbook*. 5th ed. New York, NY: McGraw-Hill Professional, 1999. Section 2.13, 11.8.
[10] ISO 9000:2005(en) Quality management systems — Fundamentals and vocabulary n.d. Iso.Org. Accessed August 9, 2021. https://www.iso.org/obp/ui/
[11] J. M. Juran, and A. Blanton Godfrey, *Juran's Quality Handbook*. 5th ed. New York, NY: McGraw-Hill Professional, 1999. Section 11.8.

[12] Contributor, Techtarget. 2019. "Quality Control (QC)." Techtarget.Com. TechTarget. December 23, 2019. https://whatis.techtarget.com/definition/quality-control-QC

[13] J. W. Senders, and N. P. Moray, *Human Error: Cause, Prediction, and Reduction*. London, England: CRC Press, p. 25, 2019.

[14] J. W. Senders, and N. P. Moray, *Human Error: Cause, Prediction, and Reduction*. London, England: CRC Press, p. 26, 2019.

[15] "Statistical Language - Types Of Errors". 2021. Abs.Gov.Au. Accessed August 2, 2021. https://www.abs.gov.au/websitedbs/D3310114.nsf/home/statistical+language+-+types+of+errors

[16] The Metrology Handbook. 2012. 2nd ed. ASQ Quality Press. Chapter 29, p. 297.

[17] H. Czichos, T. Saito, and L. E. Smith, eds., *Springer Handbook of Metrology and Testing*. 2nd ed. Berlin, Germany: Springer, p. 50, 2011.

[18] Anusha. 2018. "Types of Errors in Measurement." Electronicshub.Org. August 9, 2018. https://www.electronicshub.org/types-of-errors-in-measurement/

[19] A. Grous, *Applied Metrology for Manufacturing Engineering: Grous/Applied Metrology for Manufacturing Engineering*. London, England: ISTE Ltd and John Wiley & Sons, p. 17, 2011.

[20] J. M. Juran, and A. Blanton Godfrey, *Juran's Quality Handbook*. 5th ed. New York, NY: McGraw-Hill Professional. 23.33, p. 701, 1999.

[21] A. Grous, *Applied Metrology for Manufacturing Engineering: Grous/Applied Metrology for Manufacturing Engineering*. London, England: ISTE Ltd and John Wiley & Sons, p. 167, 2011.

[22] E. Savio, L. De Chiffre, S. Carmignato, and J. Meinertz, "Economic Benefits Of Metrology In Manufacturing," *CIRP Annals*, vol. 65, no. 1, pp. 495–498, 2016. doi:10.1016/j.cirp.2016.04.020

[23] "A Brief History of the Micrometer." 2008. Aurora, Illinois: Mitutoyo America Corporation. https://www.mitutoyo.co.jp/eng/pdf/R257_Micro.pdf, p. 8.

[24] L. K. Wee, and H. T. Ning, "Vernier Caliper And Micrometer Computer Models Using Easy Java Simulation And Its Pedagogical Design Features—Ideas For Augmenting Learning With Real Instruments," *Physics Education*, vol. 49, no. 5, pp. 493–499, 2014. doi:10.1088/0031-9120/49/5/493

[25] D. Mcarthur. 1886. MICROMETER-CALIPERS. US343478A, and issued 1886.

[26] MICROMETER. n.d. Niigata Seiki Co., Ltd. Accessed October 7, 2021. http://www.niigataseiki.net/sokutei/english/pdf/catalog/micrometer.pdf

[27] Quick Guide to Precision Measuring Instruments. 2019. Kanagawa, Japan: Mitutoyo Corporation. https://www2.mitutoyo.co.jp/eng/useful/E11003/html5.html#page=13.

[28] T. Stack, L. T. Ostrom, and C. A. Wilhelmsen, *Occupational Ergonomics: A Practical Approach*. 1st ed. Nashville, TN: John Wiley & Sons, p. 10, 2016.

[29] B. P. Bernard, 1997. "Musculoskeletal Disorders and Workplace Factors: A Critical Review of Epidemiologic Evidence for Work-Related Musculoskeletal Disorders of the Neck, Upper Extremity, and Low Back." https://www.cdc.gov/niosh/docs/97-141/pdfs/97-141.pdf?id=10.26616/NIOSHPUB97141

[30] T. Stack, L. T. Ostrom, and C. A. Wilhelmsen, *Occupational Ergonomics: A Practical Approach*. 1st ed. Nashville, TN: John Wiley & Sons, p. 305, 306, 307, 2016.

[31] J. M. F. Landsmeer, "Power Grip And Precision Handling," *Annals Of The Rheumatic Diseases*, vol. 21, no. 2, pp. 164–170, 1962. doi:10.1136/ard.21.2.164

[32] E. Y. Chao, J. D. Opgrande, and F. E. Axmear, "Three-Dimensional Force Analysis Of Finger Joints In Selected Isometric Hand Functions," *Journal Of Biomechanics*, vol. 9, no. 6, pp. 387–IN2, 1976. doi:10.1016/0021-9290(76)90116-0

[33] H. Dong, A. Barr, P. Loomer, and D. Rempel, "The Effects Of Finger Rest Positions On Hand Muscle Load And Pinch Force In Simulated Dental Hygiene Work,"

Journal Of Dental Education, vol. 69, no. 4, pp. 453–460, 2005. doi:10.1002/j.0022-0337.2005.69.4.tb03933.x

[34] W. P. Neumann, S. Kihlberg, P. Medbo, S. E. Mathiassen, and J. Winkel, "A Case Study Evaluating The Ergonomic And Productivity Impacts Of Partial Automation Strategies In The Electronics Industry," *International Journal Of Production Research*, vol. 40, no. 16, pp. 4059–4075, 2002. doi:10.1080/00207540210148862

[35] P. B. Jayakumar n.d. "There Is Significant Growth in R&D in India." Business Today. Accessed October 31, 2021. https://www.businesstoday.in/magazine/interview/story/there-is-significant-growth-in-r-and-d-in-india-297011-2021-05-26.

[36] "Innovations Moulding India's Development." n.d. Gov. In. Accessed October 31, 2021. https://msme.gov.in/innovations-moulding-indias-development

[37] S. Das. n.d. "Micro, Small and Medium Enterprises: Challenges and Way Forward." Org.In. Accessed August 2, 2021. https://rbidocs.rbi.org.in/rdocs/Speeches/PDFs/MSME6E333188172E454EBCE0461ED009C5BA.PDF

[38] H. Mift, B. Sc. 2019. "How to Use A Micrometer." Fullyinstrumented.Com. February 7, 2019. https://www.fullyinstrumented.com/how-to-use-micrometer/.

[39] https://www.amazon.in/Mitutoyo-103-137-Outside-Micrometer-0-25/dp/B00MMLIJGW/ref=pd_sbs_2/258-8513692-0335340?pd_rd_w=YgDxL&pf_rd_p=2cc6ee6d-5e48-4262-84d1-99c2a988deb6&pf_rd_r=0JSF6KSTM8HWFFVGK19G&pd_rd_r=c48acbcc-3204-41f4-a409-9a58eb15651a&pd_rd_wg=UhPFB&pd_rd_i=B00MMLIJGW&psc=1

[40] N.d. Accessed July 7, 2021. https://www.ubuy.co.in/catalog/product/view/id/20227969/s/mitutoyo-102-711-ratchet-thimble-micrometer-ratchet-thimble-0-1-quot-range-0-001-quot-graduation-0-0001-quot-accuracy

[41] R. C. Brooks, "The Development Of Micrometers In The Seventeenth, Eighteenth And Nineteenth Centuries," *Journal For The History Of Astronomy*, vol. 22, no. 2, pp. 127–173, 1991. doi:10.1177/002182869102200202

[42] "Digital Micrometer Use from Mitutoyo - How to Use a Micrometer." 2016. Youtube. November 11, 2016. https://www.youtube.com/watch?v=n0tOO0XNIptU

[43] Y. Zhang, and R. Laferriere. 2012. HIGH-SPEED MEASURINGELECTRONIC DIGITAL OUTSIDE MICROMETER. US 8,091,251 B1, and issued 2012.

[44] D. Norman, *The Design of Everyday Things*. New York, NY: Basic Books, p. 162, 2002.

[45] J. Anshel, ed., *Visual Ergonomics Handbook*. Boca Raton, FL: CRC Press, p. 107, 2005.

[46] H. You, A. Kumar, R. Young, P. Veluswamy, and D. E. Malzahn, "An Ergonomic Evaluation Of Manual Cleco Plier Designs: Effects Of Rubber Grip, Spring Recoil, And Work Surface Angle," *Applied Ergonomics*, vol. 36, no. 5, pp. 575–583, 2005. doi:10.1016/j.apergo.2005.01.014

[47] I. Dianat, M. Nedaei, and M. A. M. Nezami, "The Effects Of Tool Handle Shape On Hand Performance, Usability And Discomfort Using Masons' Trowels," *International Journal Of Industrial Ergonomics*, vol. 45, pp. 13–20, 2015. doi:10.1016/j.ergon.2014.10.006

[48] V. I. Danilyak, and V. K. Oshe, "Human Factor And Reliability Of Measuring Devices". *Measurement Techniques*, vol. 13, no. 1, pp. 121–124, 1970. doi:10.1007/bf00980478

[49] U. V. Kiran, Salapakam Renuka, Mahalakshmi Reddy, D. Kumar, "Computer Aided Prototype design of Kitchen tongs," *International Journal of Asian Regional Association for Home Economics*, vol. 17, pp. 67–72, 2010.

[50] H. Udo, T. Otani, A. Udo, and F. Yoshinaga, "An Electromyographic Study of Two Different Types of Ballpoint Pens." *Proceedings of the Human Factors and Ergonomics Society Annual Meeting*, vol. 43, pp. 491–495, 1999.

[51] MITUTOYO CORPORATION, Kawasaki-shi (JP). 2013. MICROMETER. US 2013/0091720 A1, and issued 2013.

[52] E. Henry, S. Richmond, N. Shah, and A. Sengupta. 2014. "Get a Grip: Analysis of Muscle Activity and Perceived Comfort in Using Stylus Grips of Touchscreen Tablet Computers." In. El Paso, TX, USA: The XXVIth Annual Occupational Ergonomics and Safety Conference.

[53] Mitutoyo Mfg. Co., Ltd., Tokyo, Japan. 1984. MICROMETER GAUGE. 4,485,556, and issued 1984.

[54] Mitutoyo Corporation, Kawasaki-shi (JP). 2012. HEAT INSULATING COVER AND MICROMETER. US 8.245,413 B2, and issued 2012.

[55] Mitutoyo Corporation, Kawasaki-shi (JP). 2016. ERGONOMIC MICROMETER INCLUDING TWO MODES OF ADJUSTMENT. US 9,482.509 B2, and issued 2016.

[56] H. L. Carroll, "Avoiding Parallax Error When Reading A Mercury Manometer," *Journal Of Chemical Education*, vol. 44, no. 12, p. 763, 1967. doi:10.1021/ed044p763

[57] M. Staubli, "Elimination Of Parallax Error In Haematocrit Readings Using Reflexion Haematocritometry," *Journal Of Clinical Pathology*, vol. 42, no. 8, pp. 888–889, 1989. doi:10.1136/jcp.42.8.888

Annexure

ANNEXURE 3.1   SURVEY-1 QUESTIONS (1ST PHASE)

Sec. 1

TABLE A.1

Sr. No	Survey questions
1	Name of participants
2	Age
3	Do you wear spectacles/contact lenses?
4	Education background
5	Job designation
6	Industry type
7	Company name
8	Working experience

Sᴇᴄ. 2

TABLE A 1.2

Sr. No	Survey questions
9	Which micrometers are mostly used in daily work?
11	What kind/type of micrometer do you use?
12	Have you received any formal training regarding its use?
13	How frequently do you use micrometers for measurement?
14	Do you employ any stand for micrometers stability during measurement?
15	Do you use it in a controlled temperature environment?
16	Does part of your work involve taking it to different locations?
17	Do you use any means to secure your job/object while measuring it with any instrument?

ANNEXURE 3.2   SURVEY-2 QUESTIONS (2ND PHASE) FIG. A 1.1

Study of usability of metrology instrument viz.
Micrometre screw Gauge (Mic's)

This is a survey regarding usability and ergonomic aspects of Micrometre screw gauge. Please read all the questions properly and respond accordingly. Make appropriate use of visuals given wherever needed.

1) - Discomfort Rating scale -

Please ✓ mark the level of discomfort felt by you, while using the instrument for a longer time.

0	1	2	3	4	5
Not at all	Just noticeable	Barely uncomfortable	Slightly uncomfortable	Moderately uncomfortable	Somewhat uncomfortable

6	7	8	9	10
Uncomfortable	Substantialy uncomfortable	Very uncomfortable	extensively uncomfortable	Extremely uncomfortable

Fig:1 - Reference image of wrist grasp posture for micrometer screw guage

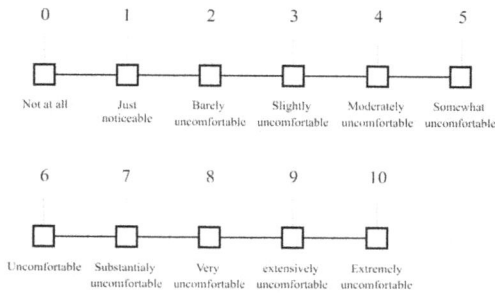

2) - Design aspect satisfaction questions -

Please ✓ mark the level of satisfaction experienced by you with regards to design aspects of current instrument design.

Physical design aspects -

- **Grasp span / area (a) -**

  Too narrow — Narrow — Satisfactory — Wide — Too wide

- **Grasp Force requirement (b) -**

  Too less — Less — Satisfactory — high — Too high

- **Frame Shape (c) -**

  Not fit at all — Poor fit — Satisfactory — Good fit — Perfect fit

- **Frame orientation (d) -**

  Too steep (less space offered) — Steep — Satisfactory — Flat — Too Flat (Much space avaialble)

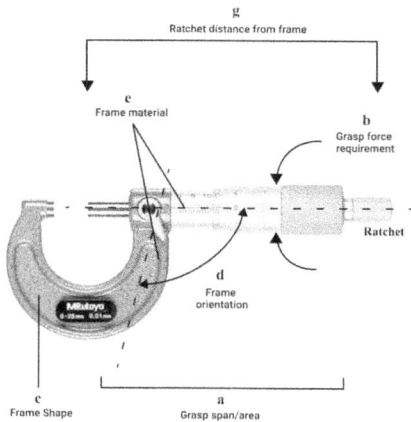

g
Ratchet distance from frame
e Frame material
b Grasp force requirement
Ratchet
d Frame orientation
c Frame Shape
a Grasp span/area

Fig:2 - Reference image (Physical design aspects of Micrometre) for the question

- **Frame material finish (e) -**

O———O———O———O———O
Too   Slippery Satisfactory Rough   Too
slippery   rough

- **Instrument weight (f) -**

O———O———O———O———O
Too   Light   Satisfactory Heavy   Too
light   heavy

- **Ratchet distance from frame (g) in case of single handed use -**

O———O———O———O———O
Too   Close   Satisfactory Far   Too
close   far

Visual design aspects -

- **Marking size (a) -**

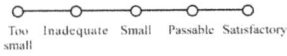

O———O———O———O———O
Too   Inadequate Small   Passable Satisfactory
small

- **Readibility (b) -**

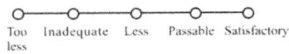

O———O———O———O———O
Too   Inadequate Less   Passable Satisfactory
less

- **Surface Finish (Reflections) (c) -**

O———O———O———O———O
Satisfactory Marginal Much   Irritating Too
                                          much

- **Marking orientation (d) -**

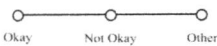

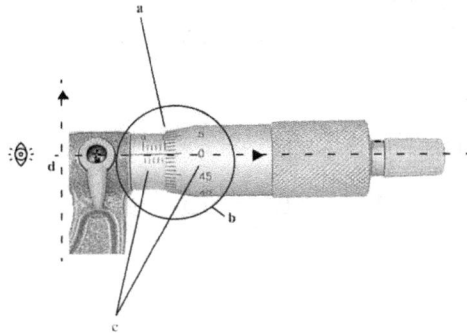

O———O———O
Okay   Not Okay   Other

Fig. 3 - Reference image (Visual design aspects of Micrometre) for the question

3) - Your hand (Palm) size -

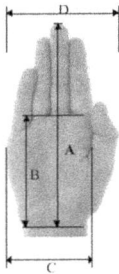

Parameter	Size in mm
A - Hand length	
B - Palm length	
C - Hand breadth without thumb	
D - Hand breadth with thumb	

4) - Your hand (Palm) size -

Tick areas marked as 1-13 (Multi choice) where you experience discomfort while using the instrument over a longer period of time

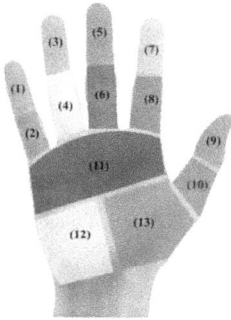

Hand and finger areas

1	LD	Little finger distal Phalanx
2	LM	Little finger middle and proximal phalanges
3	RD	Ring finger distal Phalanx
4	RM	Ring finger middle and proximal phalanges
5	MD	Middle finger distal Phalanx
6	MM	Middle finger middle and proximal phalanges
7	ID	Index finger distal Phalanx
8	IM	Index finger middle and proximal phalanges
9	TD	Thumb distal Phalanx
10	TM	Thumb proximal Phalanx
11	M	Metacarpal area
12	H	Hypothenar area
13	T	Thenar area

Hand Map (Palmer View)

Fig. 5 - Reference image (Marked sections of hand) for the question

✱ Tick ✓ the level the discomfort experienced by you between 0-10 in particular hand portions as marked by you above

			Intensity of Discomfort									
			1	2	3	4	5	6	7	8	9	10
Hand and finger areas	1	LD										
	2	LM										
	3	RD										
	4	RM										
	5	MD										
	6	MM										
	7	ID										
	8	IM										
	9	TD										
	10	TM										
	11	M										
	12	H										
	13	T										

Figure A 1.2

ANNEXURE 3.3   INTERVIEW QUESTIONS (3RD PHASE)

TABLE A 1.3

1	You seem less experienced in the field of QC, do you find it challenging?
2	How would you describe the working environment in your company?
3	Tell me something about your training and how relevant you feel it is to actual industry practices.
4	Most participants use Analogue Micrometres. Is there any reason behind it? If yes, what?
5	Why is a 1-inch micrometer mostly used in your industries?
6	It is reported that many participants make frequent use of Micrometres throughout their day. Please explain why?
7	Why do fewer people use Micrometre stands?
8	Why do fewer people wear gloves during measurement?
9	Why don't you operate a Micrometre in a controlled environment?
10	Why do you carry Micrometres with you in different locations?
11	Can you explain why it feels uncomfortable to use a Micrometre?
12	Why do you feel the grasp span is unsatisfactory?
13	Why do you feel the frame shape does not fit?
14	Why do you feel frame orientation is poor?

TABLE A 1.3 *(Continued)*

15	Why do you feel the material finish is slippery?
16	Why do you feel the ratchet distance is more from the frame?
17	Why do you feel marking orientation is not okay?
18	Why do you feel the marking size is small?
19	Why do you feel readability is poor?
20	Why do you feel reflections are more?

Index

For Product Safety Concerns and Information please contact our EU
representative GPSR@taylorandfrancis.com
Taylor & Francis Verlag GmbH, Kaufingerstraße 24, 80331 München, Germany

www.ingramcontent.com/pod-product-compliance
Lightning Source LLC
Chambersburg PA
CBHW060358220326
41598CB00023B/2958